**STEADY-STATE METHODS FOR SIMULATING
ANALOG AND MICROWAVE CIRCUITS**

THE KLUWER INTERNATIONAL SERIES
IN ENGINEERING AND COMPUTER SCIENCE
VLSI, COMPUTER ARCHITECTURE AND
DIGITAL SIGNAL PROCESSING

Consulting Editor
Jonathan Allen

Other books in the series:

STEADY-STATE METHODS FOR SIMULATING ANALOG AND MICROWAVE CIRCUITS

by

Kenneth S. Kundert
Cadence Design Systems

Jacob K. White
Massachusetts Institute of Technology

and

Alberto Sangiovanni-Vincentelli
University of California, Berkeley

KLUWER ACADEMIC PUBLISHERS
Boston/Dordrecht/London

Distributors for North America:
Kluwer Academic Publishers
101 Philip Drive
Assinippi Park
Norwell, Massachusetts 02061 USA

Distributors for all other countries:
Kluwer Academic Publishers Group
Distribution Centre
Post Office Box 322
3300 AH Dordrecht, THE NETHERLANDS

Library of Congress Cataloging-in-Publication Data

Kundert, Kenneth S.
  Steady-state methods for simulating analog and microwave circuits
/ by Kenneth S. Kundert, Jacob K. White, Alberto Sangiovanni
-Vincentelli.
      p.   cm. — (The Kluwer international series in engineering and
computer science)
    ISBN 0-7923-9069-5
    1. Microwave circuits—Mathematical models. 2. Linear integrated
circuits—Mathematical models. I. White, Jacob K. II. Sangiovanni
-Vincentelli, Alberto. III. Title. IV. Series.
TK7876.K845   1990   89-78195
621.39'5—dc20   CIP

Printed in the United States of America

Table of Contents

x

Preface

The motivation for starting the work described in this book was the interest that Hewlett-Packard's microwave circuit designers had in simulation techniques that could tackle the problem of finding steady-state solutions for nonlinear circuits, particularly circuits containing distributed elements such as transmission lines. Examining the problem of computing steady-state solutions in this context has led to a collection of novel numerical algorithms which we have gathered, along with some background material, into this book. Although we wished to appeal to as broad an audience as possible, to treat the subject in depth required maintaining a narrow focus. Our compromise was to assume that the reader is familiar with basic numerical methods, such as might be found in [dahlquist74] or [vlach83], but not assume any specialized knowledge of methods for steady-state problems.

Although we focus on algorithms for computing steady-state solutions of analog and microwave circuits, the methods herein are general in nature and may find use in other disciplines. A number of new algorithms are presented, the contributions primarily centering around new approaches to harmonic balance and mixed frequency-time methods. These methods are described, along with appropriate background material, in what we hope is a reasonably satisfying blend of theory, practice, and results. The theory is given so that the algorithms can be fully understood and their correctness established. The practical details are given so that the algorithms can be implemented easily, and results are presented to give a feel for the achievable performance. Most of the algorithms presented in this book have been implemented and tested in the *Spectre* and *Nitswit* circuit simulators.

Chapter 2 presents several typical circuits for which the steady-state response is of interest, is generally considered hard to compute, and can be computed efficiently with the specialized methods presented in this book. This chapter is intended to provide motivation for the steady-state methods presented in later chapters, as well as suggesting their application domain. The circuits are revisited in Chapter 8 to demonstrate the advantages of the steady-state methods presented in this book on practical examples. Also, various aspects of the performance of the steady-state methods are compared for each of the circuits, so as to make clear each of the method's strengths and weaknesses.

General background material is given in Chapter 3. The chapter begins with definitions of periodic, quasiperiodic, and almost-periodic signals. Generalizations of the Fourier transform suitable for almost-periodic signals are also described. Then, basic definitions and properties of differential equations in the context of initial- and boundary-value problems are presented. It is also shown how to formulate the problem of finding a periodic solution as a boundary-value problem, and the differential equations that describe analog and microwave circuits are formulated. Finally, the standard numerical algorithms for solving the initial-value problems associated with circuits are briefly described, including backward-difference formulas for numerical discretization of differential equations and Newton's method for solving nonlinear systems of algebraic equations.

In Chapter 4, additional background material is given on finite-difference and shooting methods for solving boundary-value problems. Shooting methods are covered in more detail, with both extrapolation- and Newton-Raphson-based shooting methods described. In addition, some of the difficulties of solving autonomous systems, that is oscillators, with shooting methods are presented.

Chapter 5 introduces underlying theory of harmonic balance. The harmonic balance equations are formulated and several methods for solving these equations are explored. Methods presented include minimization methods, nonlinear relaxation methods, and the Newton-Raphson method. Aspects related to implementing harmonic balance are the subject of Chapter 6. It is first shown that while in general, harmonic balance requires a quasiperiodic Fourier transform to convert signals to and from the time-domain, certain fundamental assumptions in the harmonic balance formulation allow the use of a simpler periodic Fourier transform. The several ways this impacts an implementation of harmonic balance are then described. Next, it is shown that an appropriate organization of the unknowns leads to a new technique for modifying the system Jacobian such that the resulting nonlinear relaxation method has convergence properties near that of Newton's method, but is much faster. Results from the harmonic balance program *Spectre* are used to demonstrate the effectiveness of the various techniques.

A new, mixed frequency-time (or MFT) method for computing quasiperiodic solutions directly in the time-domain are described in Chapter 7, and is compared to several other approaches. It is shown that the method is a way of formulating an approximation to the quasiperiodic requirement as a two-point boundary-value problem. The implementation of MFT in *Nitswit* is described and results from several different types of circuits are given.

Chapter 8 compares the various methods for finding steady-state solutions in both a general setting, and with regard to their performance on typical circuits. Chapter 9 summarizes and concludes the book. In the several appendices, related material is described. Appendix A lists standard nomenclature used throughout the book. In Appendix B the effects on conditioning of the placement of time and frequency points in the Almost-Periodic Fourier Transform is presented along with algorithms for minimizing the ill-conditioning. In Appendix C, the Arc-Length continuation algorithm is described as applied to solving systems of nonlinear circuit equations.

The program *Spectre* is available from the Industrial Liaison Program, Software Distribution office, Department of Electrical Engineering and Computer Science, University of California, Berkeley, CA 94720. The program *Nitswit* is available from Prof. Jacob White, Department of Electrical Engineering and Computer Science, MIT, Cambridge, Mass.

Acknowledgements

The authors wish to thank Bruce Donecker for his constant support, critical inputs and encouragements throughout this project. Greg Sorkin contributed key ideas and help in writing the sections on the APFT. James Spoto and Cadence allowed this project to be completed.

This work has been supported by Hewlett-Packard, the MICRO Program of the State of California, DARPA contracts N0039-C-87-0182 (Berkeley) and N00014-87-K-825 (MIT), an NSF Presidential Young Investigator award, and grants from IBM and Analog Devices.

STEADY-STATE METHODS FOR SIMULATING
ANALOG AND MICROWAVE CIRCUITS

Chapter 1
Introduction

Accurate computer simulation has helped reduce the cost and time required to complete many new integrated circuit designs. This is because a design can be corrected and its performance tuned faster and more economically using computer simulation than by repeated fabrication and testing of prototypes. However, to tune the performance of many analog and microwave integrated circuit designs, prototyping is still used because the commonly available computer simulation programs cannot simulate efficiently the test scenarios of interest. The main difficulty simulating general analog circuit designs is computing steady-state quantities, such as harmonic distortion, for circuits with a widely spread response spectrum. And in the specific case of microwave designs, this problem is compounded because microwave circuits typically include linear time-invariant distributed devices such as dispersive transmission lines.

The commonly used circuit simulation programs, such as SPICE [nagel75] and ASTAP [weeks73], rely on numerical integration algorithms that are well-suited for the transient analysis of nonlinear circuits, but these algorithms often cannot compute efficiently and accurately the steady-state quantities of interest to an analog circuit designer, nor can they simulate efficiently circuits containing distributed devices. Simulation programs based on phasor analysis, such as

Touchstone®[1], are commercially available, can compute easily steady-state quantities, and can include distributed devices. However, phasor analysis requires that the entire circuit be linear and therefore is not applicable to inherently nonlinear analog circuits like multipliers or mixers and, of course, the approach cannot be used to compute harmonic distortion. Providing analog or microwave circuit designers with simulation programs that can compute efficiently and accurately steady-state responses to nonlinear circuits, which often include distributed devices, is a challenging and important problem, and the subject of this book.

1. The Standard Circuit Simulation Method

The behavior of an electrical circuit constructed from resistors, capacitors, and current sources can be described with a system of equations involving node voltages, resistive currents, and capacitor charges. This system of equations can be constructed from a circuit description using nodal analysis [desoer69], which involves applying the Kirchoff Current Law (KCL) to each node in the circuit, and applying the constitutive or branch equations to each circuit element. Systems of equations so generated have the following form,

$$\frac{d}{dt}q(v(t)) = -i(v(t)) - u(t) \tag{1.1}$$

where N is the number of circuit nodes excluding the reference, $q(v(t)) \in \mathbb{R}^N$ is the vector of sums of capacitor charges at each node, $i(v(t)) \in \mathbb{R}^N$ is the vector of sums of resistor currents at each node, $u(t) \in \mathbb{R}^N$ is the vector of input currents, and $v(t) \in \mathbb{R}^N$ is the vector of node voltages.

The nodal analysis technique can be extended easily to include circuits with inductors and voltage sources by using modified nodal analysis [ho75], while still preserving the form of (1.1). The unknowns then become a mixture of node voltages and device currents, and the left-hand side of (1.1) becomes the time derivative of the

---

[1]*Touchstone* is a registered trademark of EEsof.

vector of capacitor charges and inductor fluxes.

Finding the solution to (1.1) over a specified time interval from a specified initial condition is an *initial value problem*, and computing its solution is frequently referred to as *transient analysis*. The approach to transient analysis used in programs like SPICE [nagel75] and ASTAP [weeks73] is to apply a numerical integration algorithm like the trapezoidal rule to approximate (1.1) by a sequence of implicit algebraic equations. Given a time-step, h, the trapezoidal integration algorithm applied to (1.1) yields:

$$q(v(t+h)) - q(v(t)) = -\tfrac{1}{2}h[i(v(t+h)) + u(t+h) + i(v(t)) + u(t)]$$

$$(1.2)$$

where $v(t)$ is known, and the equation must be solved to compute $v(t+h)$. Note that the time-steps are selected so that (1.2) gives a reasonable approximation to $q(v(t+h))$ given $q(v(t))$.

The iterative Newton-Raphson algorithm is usually used to solve the implicit nonlinear algebraic system given by (1.2). The Newton-Raphson iteration equation for solving a general nonlinear system of the form $F(x) = 0$ is

$$J_F(x^{(j)})(x^{(j+1)} - x^{(j)}) = -F(x^{(j)})$$

where $F(x^{(j)})$ is referred to as the Newton-Raphson residue, J_F is the Jacobian matrix of F with respect to x, and j is the iteration index.

If the Newton-Raphson algorithm is used to solve (1.2) for $v(t+h)$, the residue, $F(v^{(j)}(t+h))$, is:

$$F(v^{(j)}(t+h)) = q(v^{(j)}(t+h)) - q(v(t)) \qquad (1.3)$$

$$+ \tfrac{1}{2}h[i(v^{(j)}(t+h)) + u(t+h)) + i(v(t)) + u(t)]$$

and the Jacobian of $F(v^{(j)}(t+h))$, $J_F(v^{(j)}(t+h))$ is:

$$J_F(v^{(j)}(t+h)) = \frac{\partial q(v^{(j)}(t+h))}{\partial v} + \tfrac{1}{2}h\frac{\partial i(v^{(j)}(t+h))}{\partial v}.$$

Then $v^{(j+1)}(t+h)$ is derived from $v^{(j)}(t+h)$ by solving the linear system of equations

$$J^F(v^{(j)}(t+h))[v^{(j+1)}(t+h) - v^{(j)}(t+h)] = -F(v^{(j)}(t+h)) \qquad (1.4)$$

using some form of Gaussian elimination. The Newton iteration is continued until sufficient convergence is achieved, that is $|v^{(j+1)}(t+h) - v^{(j)}(t+h)| < \varepsilon$ and $F(v^{(j)}(t+h))$ is close enough to zero.

2. Steady State

Although most circuit simulation programs focus on transient analysis, the steady-state behavior of analog and microwave circuits is typically of primary interest to a designer. This is because certain aspects of system performance are easier to characterize and verify in steady state. Examples of quantities that are best measured when a circuit is in steady state include distortion, power, frequency, noise, and transfer characteristics such as gain and impedance. As the main subject of this book are methods for directly computing the steady state response of nonlinear and distributed circuits, before introducing these methods, we clarify what is meant by steady state.

In the most general terms, a *steady-state solution* of a differential equation is one that is asymptotically approached as the effect of the initial condition dies out. A differential equation may not have a steady-state solution, or can have any number of steady-state solutions. If there are multiple steady-state solutions, the steady-state that is asymptotically approached will depend on the initial condition. In particular, for each steady-state solution there must correspond a region of attraction for which if the initial condition is contained in the associated region, then the solution approaches the given steady-state solution. For example, consider the differential equation system that describes the circuit of two cross-coupled inverters. There is more than one steady-state solution, either one or the other inverter produces a logical one, and which will be approached depends on the initial conditions. In addition, for the case of typical inverter circuits, the region of attraction for a steady-state solution includes all initial conditions within some nonzero distance of the steady-state solution. This implies that if the solution is perturbed slightly from a given steady-state solution, it will return to that same steady-state solution. Such a steady-state solution is referred to as being *asymptotically stable*. Notice that this definition excludes lossless linear integrators and LC or harmonic oscillators from having steady-state solutions. The solutions to these circuits are not asymptotically stable.

There are several different kinds of steady-state behavior that are of interest. The first is *DC steady state*. Here the solution is an equilibrium point of the circuit and does not vary with time. Asymptotically stable linear circuits driven by sinusoidal sources eventually exhibit a *sinusoidal steady-state* solution, which is characterized as being purely sinusoidal except possibly for some DC offset. If the steady-state response of a circuit consists solely of a linear combination of a DC offset and a possibly infinite number of harmonically related sinusoids, the circuit is said to be in *periodic steady state*. Periodic steady-state results either from self oscillations or as a response to periodically varying inputs. The period of the solution is usually equal to that of the input, if it exists, though occasionally the periods of the two will be multiples of some common period. If a nonlinear circuit is driven with several periodic sources at unrelated frequencies, the circuit will typically have a steady-state response that is *quasiperiodic*. A quasiperiodic response consists of a linear combination of sinusoids at the sum and difference frequencies of a finite set of fundamental frequencies and their harmonics. The fundamental frequencies usually equal that of the input signals, though sometimes they are even multiples. Also, a fundamental frequency may result from a self oscillation rather than a stimulus. Quasiperiodic steady-state includes periodic steady state as a special case. In Figure 1.1, examples of the different types of steady-state solutions are given.

There are steady-state responses that do not fit in any of the above classifications. These occur when either the input sources are not quasiperiodic or when the circuit is strange. Examples include chaotic circuits and circuits with noise as the stimulus. In this book, we only discuss the computation of periodic and quasiperiodic solutions. In addition, most of the methods described in this book do not distinguish between solutions that are asymptotically stable and those that are not, and there are examples in which the methods described

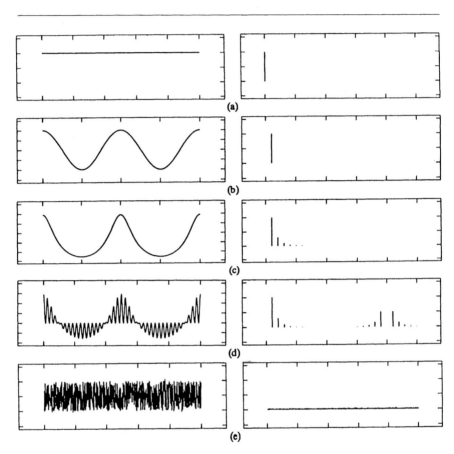

Figure 1.1: Examples of a) DC, b) sinusoidal, c) periodic, d) quasiperiodic, and e) chaotic steady state solutions in the time and frequency domains.

herein will compute periodic and quasiperiodic solutions that are not asymptotically stable. However, in most practical cases the periodic and quasiperiodic solutions found by these methods are steady-state solutions. There are methods for determining the stability of a solution, we refer the interested reader to [hente86, parker89].

3. Distributed Devices

To construct an accurate model of very high speed circuits, it is often necessary to consider the nonideal behavior of the interconnection. For example, when two distant points in a circuit are to be connected by a conductor, the conductor will be physically implemented with a strip of metal which passes over, but is insulated from, a ground plane (Figure 1.2). In microwave applications, these metal strip conductors are long enough that the voltages and currents vary appreciably over the length of the metal strip, which affects observably circuit performance. In particular, the metal strip, insulator, and ground plane form an approximation to a *transmission line*, which is the most common *distributed* device used in modeling circuits. The name *distributed* stems from the fact that the distribution of voltages and currents along the metal strip affects circuit performance.

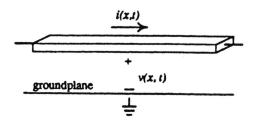

Figure 1.2: A microstrip line. (Courtesy of Steve McCormick [mccormick89]).

A transmission line differs in several ways from an ideal conductor: a signal applied at one end arrives at the other end only after some delay; if the transmission line is not properly terminated, signals will be reflected at each end and produce "ringing"; and the signal may be distorted by loss or dispersive effects due to resistance of the metal strip or nonuniformity of the insulator. The precise behavior of the transmission line can be computed by solving the wave equation system that describes the voltages and currents along the length of a transmission line,

$$L\frac{\partial i(x,t)}{\partial t} + \frac{\partial v(x,t)}{\partial x} + Ri(x,t) = 0 \qquad (1.5)$$

$$C\frac{\partial v(x,t)}{\partial t} + \frac{\partial i(x,t)}{\partial x} + Gv(x,t) = 0$$

where $v(x,t)$ and $i(x,t)$ are the metal strip voltage and current as a function of position along the strip and time; and R, L, G, and C are the series resistance and inductance, and shunt conductance and capacitance per unit length respectively. To solve for the time evolution of the currents and voltages in the transmission line, boundary conditions must be established at the ends of the transmission line, and the initial distribution of voltages and currents over the line's entire length must be known. That the need for a continuum of initial conditions, or in effect that the device has "infinite state," is another characteristic of distributed devices.

It is possible to simulate circuits with transmission lines by numerically solving (1.5) for each line. One standard numerical algorithm for solving (1.5) is equivalent to approximating the transmission line with several sections of a resistor-conductor-inductor-capacitor (RGLC) network, as in Figure 1.3, and then simulating the nondistributed, or lumped, circuit with standard circuit simulation techniques. A second approach to simulating circuits with transmission lines can be derived by exploiting the fact that only the voltages and currents at the ends, or terminals, of the line effect the rest of the circuit. For linear time-invariant transmission lines, the terminal currents can be computed from the terminal voltages by convolution with the impulse response.

That is,

$$i_j(t) = \sum_{k} \int_{-\infty}^{t} h_{jk}(t-\tau)v_j(\tau)d\tau,$$

where $i_j(t)$ and $v_j(t)$ are the jth terminal current and voltage respectively, and $h_{jk}(t)$ is the current response at terminal j due to a voltage impulse at terminal k.

Both the equivalent circuit and convolution integral approach to simulating transmission lines can be computationally expensive. In both cases, substantial additional data is used. To compute currents using the convolution integral, the voltages for all past time are needed, and in the case of a circuit equivalent, the additional devices add substantially to the number of state variables in the system. Convolution is generally the preferred time-domain method, even though the impulse responses required are difficult to compute or measure accurately. This is partly because the circuit equivalent is often hard to construct, and partly because computing a convolution is usually less expensive than simulating a circuit with the large number of additional

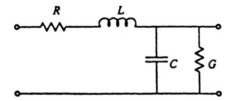

Figure 1.3: A lumped equivalent circuit to an infinitesimal section of the microstrip line of Figure 1.2. (Courtesy of Steve McCormick [mccormick89]).

state variables.

If it is possible to perform computations directly in the frequency domain, then linear time-invariant distributed devices, such as transmission lines, can be included easily. This results primarily because the frequency-domain solution consists of a sum of sinusoids, functions that are easy to manipulate analytically. Since the distributed devices are linear, superposition also allows each frequency to be handled individually. Consequently, the frequency-domain constitutive relations between currents and voltages in a distributed device are still algebraic, just like lumped (nondistributed) devices, and it is relatively easy to develop frequency-domain models for distributed devices, even those that exhibit loss, dispersion, or coupling.

As an example, consider the wave equation at the frequency ω, in which case the voltage and current take the form $V(x)e^{j\omega t}$ and $I(x)e^{j\omega t}$ where $V(x)$ and $I(x)$ are complex and are referred to as phasors.

$$\left[L\frac{\partial}{\partial t} + R\right]I(x)e^{j\omega t} + \frac{\partial}{\partial x}V(x)e^{j\omega t} = 0 \qquad (1.6\,\text{a})$$

$$\left[C\frac{\partial}{\partial t} + G\right]V(x)e^{j\omega t} + \frac{\partial}{\partial x}I(x)e^{j\omega t} = 0, \qquad (1.6\,\text{b})$$

which can be simplified to

$$[R + j\omega L\,]I(x)e^{j\omega t} + \frac{\partial}{\partial x}V(x)e^{j\omega t} = 0$$

$$[G + j\omega C\,]V(x)e^{j\omega t} + \frac{\partial}{\partial x}I(x)e^{j\omega t} = 0.$$

Notice that the time derivatives were evaluated symbolically, converting the partial differential equations to simpler ordinary differential equation in space for the phasors. To derive the constitutive relations for a lossless transmission line ($R = G = 0$) of length D, (1.6) can be solved analytically for the relationship between the phasors at one terminal, $V(0)$, $I(0)$, and those at the other, $V(D)$, and $I(D)$. In particular,

$$e^{j\omega\sqrt{LC}D}[V(D) - Z_0I(D)] = V(0) - Z_0I(0)$$

and

$$e^{j\omega\sqrt{LC}D}[V(0) + Z_0I(0)] = V(D) + Z_0I(D)$$

where $Z_0 = \sqrt{L/C}$.

4. Computing the Steady-State

Most practical analog circuits, at least ones that are reasonably close to correctly designed, will achieve a periodic or quasiperiodic steady-state from any initial condition, assuming the inputs are periodic or quasiperiodic. This presumes a reasonable circuit is "sufficiently stable," and excludes examples like lossless integrators and resonators. It is possible to calculate the steady-state responses of these "sufficiently stable" circuits using transient simulators, such as SPICE or ASTAP, by integrating numerically the differential equations that describe the circuit from some arbitrary initial condition until any transient behavior dies out. However, this fairly general approach can become impractical when the frequencies in the steady-state response are much higher than the rate at which the circuit approaches steady-state, or when the ratio of the highest to the lowest frequency present in the steady-state solution is large. In these cases, the number of discretization time-steps used by the numerical integration algorithm will be enormous, because the time interval over which the differential equations must be numerically integrated is set by the lowest frequency, or by how long the circuit takes to achieve steady-state, but size of the time-steps is constrained by the highest frequency response.

Conventional transient analysis can be accelerated by adapting it to the problem at hand. Examples of this include the methods of Gautschi and Petzold. Gautschi [gautschi61] attacked the problem of systems with lightly damped oscillatory solutions by using a custom integration method that integrates trigonometric polynomials exactly. This method allows time-steps to be used that are on the same order as the period of the oscillation. Petzold [kundert88a, petzold81] attacked the same problem by developing an integration method that follows the envelope of the solution rather than the solution itself.

There are a wide variety of methods that directly compute the steady-state solution more efficiently than integrating numerically the differential equations that describe a circuit from some arbitrary initial condition. In the time-domain, it is possible to derive faster algorithms if the steady-state problem is recast to one of finding an initial condition for which the solution to the differential equation is immediately the steady-state. It is also possible to derive frequency-domain methods for computing the steady-state solution, in which case what is calculated are the coefficients for truncated trigonometric (sine and cosine) series expansions of the steady-state solution. Frequency-domain methods have an advantage over time-domain methods in that they can incorporate distributed devices easily, but if the problem is very nonlinear, the large number of terms required in the trigonometric-series expansions can make frequency-domain methods inefficient.

4.1. Time-Domain Methods

Finding the periodic steady-state solution of a circuit involves finding the initial condition for the circuit's associated differential equation system such that the solution at the end of one period matches the initial condition. Problems of this form, that of finding a solution to a system of ordinary differential equations that satisfies constraints at two or more distinct points in time, are referred to as *boundary-value problems*. For example, finding the solution to

$$\dot{x}(t) = f(x(t), t)$$

over the interval $[0, T]$ is a two-point boundary-value problem if the solution is required to satisfy

$$g(x(0), x(T)) = \mathbf{0}.$$

As mentioned above, it is possible to solve the boundary-value problem associated with computing the steady-state of a well-behaved circuit by solving the initial-value problem over a sufficiently long time interval. For most cases, there are more efficient methods that solve the boundary-value problem more directly. However, the numerical methods used for direct solution of boundary-value problems are innately more complicated than those used to solve initial value problems. In an initial-value problem it is possible to compute the solution by propagating forward from some known initial condition, without concern for the unconstrained final condition. This is not the case for a two-point boundary-value problem. Not only is the initial condition unknown *a priori*, but the initial and final condition must satisfy some constraint, and this implies that the numerical calculation must propagate information forward, from the initial condition, and backward, from the final condition.

There is considerable literature on the subject of computing the solution to boundary-value problems. For example, see [childs79] [fox57] [hall76] [keller68] [keller76] [press86] [stoer80]. The methods most commonly applied to circuit problems are shooting methods [aprille72b, skelboe80]. Shooting methods solve boundary-value problems by computing the solution to a succession of initial value problems, with steadily improved guesses at an initial condition which results in steady-state. For example, the shooting algorithm for computing a circuit's periodic steady-state begins by simulating the circuit for one period using some guessed initial condition. The computed solution at the end of the period is then checked, and if it does not match the initial condition, the initial condition is adjusted. The circuit is then resimulated, with the adjusted initial condition, and this process is repeated until the initial condition and the solution after one period match closely enough. How the initial condition is adjusted on each iteration depends on the particular variant of shooting method.

The other standard approach for solving boundary-value problems is the finite-difference method. Methods of this type apply a finite-difference approximation to the system of differential equations to generate a sequence of algebraic equations. The sequence of algebraic equations are then solved simultaneously with the boundary constraints to compute the boundary-value problem solution directly. Finite-difference methods are not commonly used for circuit problems because the size of the system that must be solved is very large. In particular, the number of unknowns is equal to the product of the number of state variables and the number of time-steps.

The computational effort of shooting and finite-difference methods for calculating steady-state solutions of a circuit is significantly larger than that of solving an initial value problem over the same interval, but the computation time required is mostly independent of any circuit time constants or frequency separations. Thus, methods based on integrating numerically from an arbitrary initial condition are often the most efficient, if the circuit does not have slow time constants or widely-separated frequencies. However, for any circuit there is always some value for the time constants or frequencies for which shooting or finite-difference methods are more efficient, and in practice this difference can often be more than an order of magnitude.

4.2. Frequency Domain Methods

The time-domain discretization techniques described above for computing a circuit's steady-state solution can be interpreted as representing the solution with low-order piecewise polynomials which are connected at the time-step locations. If the steady-state is nearly sinusoidal, which is common for many analog circuits, a large number of these polynomial pieces, or equivalently a large number of time-steps, will be required to represent the solution accurately. Alternatively, the steady-state solution can be represented with a trigonometric (sine and cosine) series. The advantage of the trigonometric-series approach is that the steady-state solution can often be very accurately represented with a few terms. For example, if the input is sinusoidal and the circuit is linear, only two terms of the trigonometric series (one sine and one cosine) will represent the solution exactly, whereas a piecewise polynomial representation will only approximate the solution no matter how many terms are used.

The advantage of the trigonometric-series approach is also particularly compelling in the case where the steady-state solution is quasi-periodic with very widely spaced fundamental frequencies. For example, if the steady-state solution is precisely the sum of two sinusoids which differ in frequency by a factor of 100, the trigonometric-series representation will have only four terms and be exact, but much more than one hundred time-points will be needed for the piecewise-polynomial representation. And even then, the piecewise-polynomial representation will only be approximate.

The most commonly used frequency-domain method for circuits is to calculate the steady-state solution by directly computing with the coefficients of the trigonometric-series representation. If the circuit is linear, and the input sources are sines and cosines, computing directly with the coefficients is equivalent to phasor analysis [desoer69]. In general, the advantage of working directly with the coefficients is that linear dynamic operations like differentiation and integration become simple algebraic operations such as multiplying or dividing by frequency. This implies that for analyzing linear time-invariant circuit devices, the coefficients of the response are easily calculated by exploiting superposition and using phasor analysis on the device. That is, the coefficients of the response at a given frequency of a linear time-invariant device are computed by determining the product of the magnitude and phase of the device's response at that frequency by the magnitude and phase represented by the stimulus coefficients at *only* that frequency. The calculations are trivial for any lumped components such as resistors, capacitors, inductors, and is also simple for any distributed component that can be described by a frequency response. In particular, transmission lines, even ones that exhibit dispersion, can be included much more easily that in time-domain simulators.

Computing the coefficients of the response of nonlinear devices is more difficult than for linear devices, in part because superposition no longer applies, and each of the coefficients of the response can be complicated functions of all the coefficients of the stimulus. And in addition, for an arbitrary nonlinearity, the coefficients of the response cannot be computed directly from the coefficients of the stimulus, although it is possible for some useful special cases [steer83]. The response coefficients can be calculated approximately by converting into the time domain and then back again. Specifically, the inverse Fourier transform is used to convert the coefficient representation of the stimulus into a sampled-data representation. In the sampled-data representation the nonlinear devices can be evaluated easily and then the results can be converted back into response coefficients using the forward Fourier transform.

Frequency-domain methods applied to nonlinear circuits where the computations are performed using the trigonometric-series coefficients directly are referred to as harmonic-balance methods, mostly for historical reasons. The name stems from an approach based on balancing of currents between the linear and nonlinear subcircuits. Working with the coefficients in the harmonic balance method in effect replaces the nonlinear integro-differential equations that describe a circuit by a system of algebraic nonlinear equations in terms of the coefficients, whose solution is an approximation to the steady-state response of the circuit.

Chapter 2
Motivation

Using numerical integration to compute a circuit's transient response is by far the most commonly performed simulation of nonlinear circuits, even when the user is interested in steady-state behavior. As motivation for the chapters that follow, this section presents examples from several general classes of analog and microwave circuits for which the standard performance criteria must be measured when the circuit is in steady-state. And, to be honest, the circuits selected are ones for which the algorithms presented in the rest of this book are substantially more efficient and / or more accurate than standard techniques for computing steady-state. The circuits of this section are revisited in Chapter 8 at the end of this book, and used as examples to show the strengths and weaknesses of the methods presented in the following chapters.

1. Self-Biasing Amplifiers

Self-biasing amplifiers are AC-coupled amplifiers that use a simple biasing network to reduce the gain of an amplifier when very large signals are applied. Consider the self-biasing JFET amplifier shown in Figure 2.1. When a small input signal is applied, the DC level measured at the transistor's gate is set by the resistor R_{bias} and the current characteristic of the gate-source junction diode. However, when large input signals are applied, the gate-source junction diode of the FET will conduct more strongly on the positive peaks than for the negative peaks, which results in an increase in DC current flowing into the gate.

No DC current can flow though a capacitor, and so the DC voltage on the gate must fall so that R_{bias} will provide an average DC current equal to that flowing into the gate. When the DC bias level of the gate falls, so does the gain of the amplifier.

Computing this circuit's transient behavior until steady-state is achieved is expensive for two reasons. First, the time constant of the bias circuitry is always chosen to be much longer than the period of the lowest input frequency. Thus, when simulating a wide bandwidth amplifier at its highest input frequencies, the time required for the bias network to settle could be on the order of thousands of cycles of the input, making a transient analysis very expensive. Second, the time required for the bias circuitry to settle is difficult to predict because it is mostly determined by the nonlinearity in the gate-source junction diode, and is also often so slow that most tests for detecting when steady-state has been reached are unreliable. This requires a conservative user to simulate the circuit over a much longer time interval than is probably necessary. The other alternative is to trust a suspect answer with no way of checking its accuracy, except to resimulate the

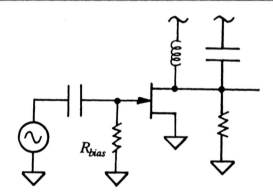

Figure 2.1 : A self-biasing amplifier.

circuit over a longer time interval to see if the answer changes.

2. Mixers

A mixer is a nonlinear circuit commonly used in communications to translate a signal from one frequency to another. A mixer has two inputs, one usually referred to as (RF), for Radio Frequency, which is to be translated, and the other (LO), for Local Oscillator, which performs the translation. The output predominantly contains two signals, one at the sum of the RF and LO frequencies, and one at their difference. Usually only one of the signals is desired, and so the mixer is followed by a filter. A simple double-balanced mixer is shown in Figure 2.2.

Mixers are very difficult to simulate for two reasons. First, the frequencies of the signals present can be very widely separated. Second, the settling time of the output filter can be very much longer than the period of the lowest frequency present in the mixer. Consider the down conversion mixer in the HP8505 network analyzer [dalichow76]. The mixer has an input RF frequency that ranges from 500kHz to 1.3GHz and an LO frequency that is always offset from the RF by 100kHz. The desired output frequency is 100kHz, and the output is fed directly into a high-Q low-pass filter to assure that this is the only signal present at the output. Simulating this circuit is extremely difficult because the ratio of the input to output frequency can be as high as 13,000 to 1, and because the output filter has a long settling time. Transient simulation of this circuit requires time-steps much less than 1 ns to capture the high frequencies, and a simulation interval of at least 100μ to capture ten periods of the filter output — a minimum of 10^5 time-points are needed. It is difficult to present meaningful results in the presence of such a large amount of data, particularly with the vastly different time scales involved. Normally, this problem is avoided by converting the solution into the frequency domain, but the many unequally spaced time-points generated by the simulator along with the nonperiodic signals make this a difficult task.

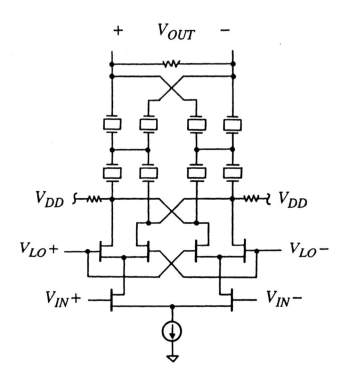

Figure 2.2 : GaAs double balanced mixer with crystal lattice filter.

3. Narrow-Band Amplifiers and Filters

Finding the periodic steady-state response of a narrow-band amplifier and / or filter can be expensive using transient analysis because the set-tling time of the amplifier is usually long in comparison to the period of its center frequency. Computing quantities like intermodulation dis-tortion, however, can be *extremely* expensive. The method used to measure standard distortion of a wide-band amplifier is to apply a pure sinusoid to the input, and determine by how much the output deviates from being a pure sinusoid in steady state. The harmonic distortion products in the output signal will fall at frequencies that are integer multiples of the input frequency. If this same technique were applied to measure the distortion of a narrow-band amplifier, the distortion products would be attenuated because they are outside the bandwidth of the amplifier, and the calculated amount of distortion would be much too low.

Instead, distortion is measured in narrow-band amplifiers by applying two pure sinusoids with frequencies well within the bandwidth of the amplifier (call these frequencies f_1 and f_2). The harmonics of these two frequencies will be outside the bandwidth of the amplifier, however there will be distortion products that fall at the frequencies $2f_1 - f_2$ and $f_1 - 2f_2$. These frequencies should also be well within the bandwidth of the amplifier and so can be used to measure accurately what is referred to as the intermodulation distortion produced by the amplifier.

Computing the intermodulation distortion using transient analysis requires simulation over a time interval greater than several periods of the difference frequency, $f_1 - f_2$. If numerical integration is used, the maximum time-step used must be smaller than the minimum period of f_1 or f_2. And as f_1 and f_2 must be close so that the distortion harmonics are in the passband, the ratio of the simulation interval to the maximum usable time-step is very large.

4. Low-Distortion Amplifiers

Low distortion amplifiers are found in demanding applications were linearity is of crucial importance. Examples of such applications include high resolution data-acquisition systems and high-fidelity audio equipment. When driven by a sinusoid, the distortion products generated by these amplifiers is often required to be 80–120 dB ($10^4 - 10^6$) below the fundamental. In order to measure this level of detail accurately, a transient simulator must place its time-points very closely. It also must compute the time-domain response until the initial transient has decayed to a level well below the expected distortion products. Both of these contribute to the expense of the simulation. As a practical matter, many transient simulators use algorithms that generate small amounts of random error that can completely mask small distortion products. An example is the bypass algorithms that monitor the voltages on the terminals of a device and use currents from a previous iteration or time-step if the voltages have not changed much. As devices go out of bypass, a small jump discontinuity occurs in the current. These discontinuities create a "noise floor" that conceals small signals.

5. Switched-Capacitor Filters

Switched-capacitor filters are simply active RC filters where the resistors have been replaced with a switch and a capacitor. The switch and capacitor are configured as shown in Figure 2.3. The switch alternates between its two positions at the clock frequency, and during each cycle transfers $C(V_a - V_b)$ coulombs of charge from terminals **a** to **b**. Thus, if the clock rate is much higher than the highest frequency present in $V_a(t)$ and $V_b(t)$, then the average current transferred equals $Cf_c(V_a - V_b)$, where f_c is the frequency of the clock. For low frequency signals, the switch and capacitor act like a resistor with resistance $R = Cf_c$.

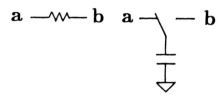

Figure 2.3 : A resistor and its switched-capacitor equivalent.

That the clock frequency must be considerably higher than the frequency of the signals being filtered make the transient analysis of switched-capacitor filters particularly expensive. The problem is aggravated because switched-capacitor filters are large circuits. Designers have avoided the expense of detailed simulation of these circuits by replacing the op amps in the active filters with algebraic macromodels and the MOSFET switches with ideal switches. For the actual simulation, they formulate and solve the charge equations once per clock transition [de man80, fang83].

This approach allows the design of a switched-capacitor filter to be verified from a high-level point of view (i.e., the circuit topology, clocking scheme, and capacitor values are all verified), but error mechanisms due to switch and op-amp dynamics are ignored completely. These effects are quite important and can only be explored using detailed circuit simulation. For example, consider the effect of a power dissipation specification on filter distortion. Circuit designers usually fix the clock rate of the filter based on external specifications and then reduce the bias current in the op-amps as much as possible. As bias current is reduced, so is the speed of the op-amps. The designers must be careful not to reduce the bias current so low as to interfere with the op-amps ability to handle the required clock frequency. When the op-amp / switch system is not able to settle during a clock cycle, distortion increases dramatically. The macromodel / charge redistribution approach to simulating switched-capacitor filters is useless when trying to predict the lowest possible op-amp bias current because of the necessary assumption that all signals reach their equilibrium values before each clock transition.

6. Traveling-Wave Amplifiers

A traveling-wave or distributed amplifier is a circuit that is very important at microwave frequencies. An example amplifier is shown in Figure 2.4 where the small rectangles represent microstrip transmission lines. Traveling-wave amplifiers (TWAs) are representative of most microwave circuits in that they contain a large number of distributed components. These components are difficult to simulate numerically

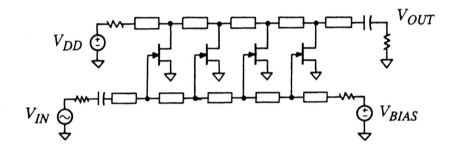

Figure 2.4: A four-segment GaAs traveling-wave amplifier.

with standard transient analysis techniques for two reasons. First, formulating the model equations for nonideal distributed components such
as lossy or dispersive transmission lines involves determining the
transmission line impulse responses. Second, evaluating the transmission line during simulation involves calculating a convolution integral
over all past data.

7. Measured Devices

Many of the linear distributed devices used at microwave frequencies
are varied and complex enough that it is not possible to develop adequate analytic models for each kind. In these cases, designers rely on
measured data to describe their devices to a simulator. Network
analyzers which, for many reasons, make measurements in the frequency domain, are used to characterize these linear distributed devices
at microwave frequencies.

In order to include a frequency domain description of a device
into a time domain simulator, the frequency response must be
transformed into an impulse response. If the impulse response does

not contain singularities, it can be computed using the discrete Fourier transform directly. If not, the approximate transformation can be quite inaccurate if not carefully computed, and precisely how to do this is an open question.

8. Crystal and Cavity Oscillators

Crystal and cavity oscillators use very high-Q resonators to achieve very high stability and low noise. The Q of the best resonators can be as high as 1,000,000, leading to a Q in well designed oscillators of up to 100,000. A Q of 100,000 implies that the time constant of the turn-on transient for an oscillator with such a Q is roughly 100,000 cycles of the oscillation frequency in length. Clearly, transient simulation of such circuit to steady-state is very painful, however steady state is required in order to determine important characteristics, such as the output power and frequency of the oscillator. Designers try to reduce the time required to simulate these circuits by carefully choosing the initial state of the resonator to eliminate any transients. However, in this example, it would be necessary to simulate the oscillator for 1000 cycles simply to notice a 1% difference in the signal envelope. Designers usually do not have the patience to find the initial state that results immediately in steady-state when to do so requires the trial-and-error selection of initial states and where each guess requires simulating the oscillator for over 1000 cycles. Thus, designers usually settle for approximate hand calculations when determining the steady-state characteristics of their oscillators.

Chapter 3
Background

This chapter both defines terminology to be used through the rest of the book and presents basic material on several topics including: the Fourier representation of signals, system and canonical formulation, and numerical methods for solving initial value and nonlinear algebraic problems. Standard notation is used throughout, although a complete list of nomenclature is include in Appendix A.

1. Fourier Representation of Signals

A *signal* is a function that maps either $\mathbb{R}$ (the reals) or $\mathbb{Z}$ (the integers) into $\mathbb{R}$ or $\mathbb{C}$ (the space of real pairs)[1]. The domain and range of the map are physical quantities; the domain is typically time or frequency, and the range is typically voltage or current. A signal whose domain is time is called a *waveform*; one whose domain is frequency is called a *spectrum*. All waveforms are assumed $\mathbb{R}$-valued whereas all spectra are assumed $\mathbb{C}$-valued.

A waveform x is *periodic* with *period T* if $x(t) = x(t+T)$ for all t. We let $P(T;E)$ denote the space of periodic functions of finite

---

[1]Throughout this book, the trigonometric Fourier series is used rather than the exponential to avoid problems with complex numbers and nonanalytic functions when deriving the harmonic Newton algorithm. Thus, a signal at one frequency in a spectrum is described using the coefficients of sine and cosine. The pair of these are said to reside in $\mathbf{C} = \mathbb{R}^2$ rather than $\mathbb{C}$.

power with period T on the domain E that can be represented by the sum of T-periodic sinusoids. Specifically, $P(T;\mathbb{R})$ consists of waveforms for which

$$x(t) = \sum_{k=0}^{\infty} (X^C(k)\cos\omega_k t + X^S(k)\sin\omega_k t), \qquad (3.1)$$

where $X^C(k), X^S(k) \in \mathbb{R}$, $\omega_k = 2\pi k/T$, and,

$$\sum_{k=0}^{\infty} [(X^C(k))^2 + (X^S(k))^2] < \infty. \qquad (3.2)$$

A waveform is *almost periodic* if it can be represented by the sum of at most a countable number of sinusoids [hale80] [corduneanu68] [besicovitch32] [bohr47] (here there is no assumed relationship between the frequencies of the sinusoids). We use $AP(\Lambda;E)$ to denote the space of all almost-periodic waveforms of finite power on domain E over the set of frequencies Λ. Thus, $AP(\Lambda;\mathbb{R})$ consists of waveforms for which

$$x(t) = \sum_{\omega_k \in \Lambda} (X^C(k)\cos\omega_k t + X^S(k)\sin\omega_k t), \qquad (3.3)$$

where $\Lambda = \{\omega_0, \omega_1, \omega_2, \cdots\}$ and (3.2) is satisfied. If Λ is finite with K elements, it is denoted Λ_K. An evenly sampled almost-periodic signal is denoted $AP(\Lambda;Z(T,\theta))$ where Z is the set of sample times $Z(T,\theta) = \{t : t = sT + \theta, s \in Z\}$. Thus $x \in AP(\Lambda;Z(T,\theta))$ implies

$$x(s) = \sum_{\omega_k \in \Lambda} (X^C(k)\cos\omega_k(sT + \theta) + X^S(k)\sin\omega_k(sT + \theta)),$$

If there is a set of d frequencies $\{\lambda_1, \lambda_2, \ldots, \lambda_d\}$ and Λ is such that

$$\Lambda = \{\omega : \omega = k_1\lambda_1 + k_2\lambda_2 + \cdots + k_d\lambda_d; k_1, k_2, \ldots, k_d \in Z\}$$

$$(3.4)$$

then Λ is a *module*[2] of dimension d and the frequencies $\{\lambda_1, \lambda_2, \ldots, \lambda_d\}$ are referred to as the *fundamental frequencies* and form a basis (called the fundamental basis) for Λ. For each $\omega \in \Lambda$ to correspond uniquely to a sequence of harmonic indices $\{k_j\}$, the sequence of fundamental frequencies $\{\lambda_j\}$ must be linearly independent over the rationals (that is $\sum_{j=1}^{d} k_j \lambda_j = 0$ implies $k_1 = k_2 = \cdots = k_d = 0$). If Λ is a module, then $AP(\Lambda; E)$ is also denoted $QP(\lambda_1, \lambda_2, \ldots, \lambda_d; E)$. Waveforms belonging to such a set are referred to as d-fundamental *quasiperiodic* or simply *d-quasiperiodic*. Note that $P(T; \mathbb{R}) = QP(\lambda_1; \mathbb{R})$ if $\lambda_1 = 2\pi/T$, and $P(T; \mathbb{R}) \subset QP(\lambda_1, \lambda_2, \ldots, \lambda_d; \mathbb{R})$ if for some j, $\lambda_j = 2\pi/T$.

The pair $X(k) = [X^C(k) \ X^S(k)]^T \in \mathbb{C}$ is the *Fourier coefficient* of the *Fourier exponent* ω_k and $X = [X(0), X(1), X(2), \cdots]^T$ is called the frequency-domain representation, or *spectrum*, of x. Conversely, x is the time-domain representation, or *waveform*, of X. If x is almost periodic and all the frequencies $\omega_k \in \Lambda$ are distinct, (i.e., $\omega_i \neq \omega_j$ for all $i \neq j$) then there exists a linear invertible operator, $\mathbb{F}$ referred to as the Fourier operator, that maps x to X.

1.1. Truncation and Discretization

In order for operations involving quasiperiodic signals to be computationally tractable, it is necessary to truncate the frequencies to a finite set. When stimulating a circuit at d fundamental frequencies, the circuit responds in steady state (if such a solution exists) at frequencies equal to the sum and difference of the fundamental frequencies and their harmonics. There are two popular methods for truncating the set of frequencies, the box and diamond truncations. With the box truncation, only the first H harmonics of each fundamental are considered:

---

[2]Roughly, a module is a set with an identity that is closed under vector addition and scalar multiplication. In this module, the vectors are real numbers, the scalars are integers and the identity is zero.

$$\Lambda_K = \{\omega : \omega = k_1\lambda_1 + k_2\lambda_2 + \cdots + k_d\lambda_d; \tag{3.5}$$

$$|k_j| = 0, 1, \ldots, H; \text{ for } 1 \le j \le d; \text{ first nonzero } k_j \text{ positive}\}$$

where $K = \frac{1}{2}((2H + 1)^d + 1)$. The first nonzero k_j must be positive to eliminate frequencies from Λ_K that are negatives of each other. When there are two fundamentals $(d = 2)$, this truncation results in a square grid of frequency indices as illustrated in Figure 3.1a.

The diamond truncation limits the absolute sum of the indices k_j to be less than or equal to H:

$$\Lambda_K = \{\omega : \omega = k_1\lambda_1 + k_2\lambda_2 + \cdots + k_d\lambda_d; \tag{3.6}$$

$$k_j \in \mathbf{Z}; \sum_{j=1}^{d} |k_j| \le H; \text{ first nonzero } k_j \text{ positive}\}$$

where $K \approx \dfrac{2^{d-1}H^d}{d!}$. For $d = 2$, $K = H^2 + H + 1$. When there are two fundamentals, this truncation produces a "diamond" grid as shown in Figure 3.1b. Other truncation schemes are certainly possible. If much is know *a priori* about the steady-state waveforms, this information can be used to chose an appropriate truncation scheme that will substantially improve the efficiency and accuracy of a frequency domain simulation method.

If a finite set of frequencies Λ_K are being considered, it is often useful to consider generating the set from fundamental frequencies that are not linearly independent over the rationals, as long as each $\omega_k \in \Lambda_K$ still corresponds uniquely to a valid sequence of harmonic indices $\{k_j\}$. This is particularly useful for systems forced by well separated but harmonically related inputs. For example, if a circuit is excited by both a 1 and a 100 Hertz signal, most of the signal energy will be in first few harmonics of 1 Hertz, the first few harmonics of 100 Hertz, and their sums and differences. It would be inefficient, although correct, to consider all the frequencies as harmonics of 1 Hertz as more than 100 harmonics would be needed to capture most of the signal energy.

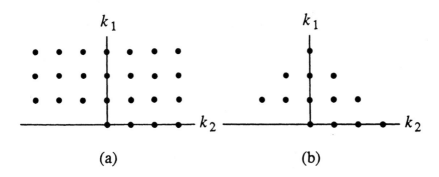

Figure 3.1: Two different ways of truncating the set of frequencies to be finite. The box (a) and diamond (b) truncations.

Once Λ has been truncated to some finite subset Λ_K, it is possible to discretize the waveforms, or represent them as sequences of finite length. Assuming that $\omega_0 = 0 \in \Lambda_K$ and $x \in AP(\Lambda_K; \mathbb{R})$, then x is uniquely specified by almost any set of $S = 2K - 1$ samples, that is, a set $\{x(t_i) : i = 1, 2, \ldots, S; t_i \neq t_j$ for $i \neq j\}$. This done, the Fourier operator becomes a finite-dimensional operator that depends both on Λ_K and the S time-points used to sample the waveform. Once the fundamental frequencies and the truncation scheme are specified, Λ_K is fixed, but we are free to choose the time-points as we see fit with the one constraint that $\mathbb{F}$ be invertible.

2. Almost-Periodic Fourier Transform

This section presents a numeric Fourier transform that is an extension over the standard Discrete Fourier Transform (DFT) in the sense that the restriction of harmonically related frequencies is removed. This more general transform is referred to as the Almost-Periodic Fourier

Transform (APFT).

We now consider deriving $\mathbb{F}$ in the case where the spectrum has finite number of frequencies, which for convenience we denote by $\Lambda_K = \{\omega_0, \omega_1, \omega_2, \ldots, \omega_{k-1}\}$ and assume all frequencies are distinct ($\omega_j \neq \omega_k$ when $j \neq k$) and that $\omega_0 = 0$. As the spectrum is finite, it is possible to sample a waveform at a finite number of time-points and calculate its Fourier coefficients. Since both spaces involved are finite dimensional and $\mathbb{F}$ is linear, $\mathbb{F}$ and its inverse $\mathbb{F}^{-1}$ can be viewed as matrices acting on the vectors of samples and coefficients, respectively. The finite-dimensional operator that relates a finite spectrum to a finite sequence of sample points of the waveform denoted

$$\sum_{\omega_k \in \Lambda_K} (X^C(k)\cos\omega_k t + X^S(k)\sin\omega_k t) = x(t)$$

is derived by sampling the waveform at the chosen S time-points. This results in the set of S equations and $2K - 1$ unknowns

$$\Gamma^{-1}\begin{bmatrix} X(0) \\ X^C(k) \\ X^S(k) \\ \vdots \\ X^C(K-1) \\ X^S(K-1) \end{bmatrix} = \begin{bmatrix} x(t_1) \\ x(t_2) \\ x(t_3) \\ \vdots \\ x(t_S) \end{bmatrix} \tag{3.7 a}$$

where

$$\Gamma^{-1} = \begin{bmatrix} 1 & \cos\omega_1 t_1 & \sin\omega_1 t_1 & \cdots & \cos\omega_{K-1} t_1 & \sin\omega_{K-1} t_1 \\ 1 & \cos\omega_1 t_2 & \sin\omega_1 t_2 & \cdots & \cos\omega_{K-1} t_2 & \sin\omega_{K-1} t_2 \\ 1 & \cos\omega_1 t_3 & \sin\omega_1 t_3 & \cdots & \cos\omega_{K-1} t_3 & \sin\omega_{K-1} t_3 \\ \vdots & \vdots & \vdots & & \vdots & \vdots \\ 1 & \cos\omega_1 t_S & \sin\omega_1 t_S & \cdots & \cos\omega_{K-1} t_S & \sin\omega_{K-1} t_S \end{bmatrix} \tag{3.7 b}$$

If the frequencies ω_k are distinct, and if $S = 2K - 1$, this system is invertible for almost all choices of time-points, and can be compactly written as $\Gamma^{-1}X = x$. Inverting Γ^{-1} gives $\Gamma x = X$. Γ and Γ^{-1} are a discrete Fourier transform pair.

Given a finite set Λ_K of distinct frequencies, and a set of time-points, we say that Γ and Γ^{-1} are one implementation of the almost-periodic Fourier transform for $AP(\Lambda_K; \mathbb{R})$. Once Γ and Γ^{-1} are known, performing either the forward (using Γ) or inverse (using Γ^{-1}) transform requires just a matrix multiply, or $(2K-1)^2$ operations; this is the same number of operations required by the DFT.

The DFT is a special case of (3.7) with $\omega_k = k\omega$ $(k = 0, 1, 2, \ldots, K-1)$ and $t_s = sT/S$ $(s = 1, 2, \ldots, S)$, i.e. when the frequencies are all multiples of a single fundamental and the time-points are chosen equally spaced within the period. The DFT and its inverse, the IDFT, have the desirable property of being well conditioned, which is to say that very little error is generated when transforming between x and X. From the matrix viewpoint, the high accuracy of the DFT corresponds to the fact that the rows of Γ^{-1} are orthogonal. (More is said about this in Appendix B.) Unfortunately, the DFT and the IDFT are defined only for periodic signals.

For almost-periodic signals, if the time-points are not chosen carefully, Γ^{-1} can be ill-conditioned. In particular, choosing time-points to be equally spaced often is a bad strategy when signals are not periodic. Unlike the periodic case, it is in general impossible to choose a set of time-points over which the sampled sinusoids at frequencies in Λ_K are orthogonal. In fact, it is common for evenly sampled sinusoids at two or more frequencies to be nearly linearly dependent, which causes the ill-conditioning problems encountered in practice. This ill-conditioning can greatly magnify aliasing. Thus, it is important to choose a set of time-points that results in well-conditioned transform matrices. This topic is considered further in Appendix B.

3. Systems

A collection of devices is called a *system* if the devices are arranged to operate on input signals (the *stimulus*) to produce output signals (the *response*). For a given solution of a system's describing equations, the system is considered to be in *steady state* if the solution is *asymptotically stable.* By this it is meant that any solution near the steady-state solution asymptotically approaches it as time increases [hirsch70]. This book considers only systems in periodic and quasi-periodic steady state.

A system is *autonomous* if both it and its stimulus are time invariant, otherwise it is *forced.* An oscillator is an example of an autonomous system while an amplifier, a filter, and a mixer are usually forced systems. An *algebraic* or *memoryless* device or system is one whose response is only a function of the present value of its stimulus, not past or future values. Traditionally, a lumped device is one whose physical dimensions are much smaller than the wavelengths of the signals present in the circuit. In this book, a device is considered *lumped* if it is accurately modeled with an algebraic function of a finite number of network variables (voltage, current, charge, flux). Any device that is not lumped is *distributed.* A lumped system contains only lumped devices.

3.1. State

A set of data qualifies to be called the *state* of a system if it satisfies the following two conditions [desoer69].

1. For any time, say t_1, the state at t_1 and the stimulus specified from t_1 on, determine uniquely the state at any time $t > t_1$.

2. The combination of the state and the stimulus at any time t determine uniquely the value at time t of any significant quantity of the system. If the system is an electrical circuit, the significant quantities are usually considered the network variables (v, i, q, ϕ).

In general the state is not unique. There may be many collections of network variables that qualify as the state.

For lumped systems, the state is a finite set of numbers, usually arranged as a vector. The components of this *state vector* are referred to as *state variables.* The state of a distributed device is described using one or more waveforms rather than a finite set of numbers. For example, consider an ideal transmission line. At some time t_1, one must know the voltage and current waveforms along the entire length of the line as well as the stimulus from t_1 on in order to predict accurately, say, the future voltages at the ends of the line.

A system can be described with an equation of the form

$$f(x,t) = 0 \qquad (3.8)$$

where $x(t)$ is a vector of network variables and f is a general function and can include integration and differentiation. If it is possible to reformulate this equation into the form

$$\dot{y}(t) = g(y(t),t) \qquad (3.9)$$

where the $y(t)$ is a vector of state variables and if there is a one-to-one correspondence between every solution y of (3.9) and every solution x of (3.8), then (3.9) is a *state equation in normal form* for the system described by (3.8). A solution x *of (3.8) is called a trajectory* and the corresponding y *is the state trajectory.*

Define ϕ to be the function that maps the state y_0 at t_0 into the solution to 3.9. That is,

$$y(t) = \phi(y_0, t_0, t). \qquad (3.10)$$

This function is referred to as the *state-transition function*. It has the characteristic that if $y(t_1) = \phi(y(t_0), t_0, t_1)$ and $t_1 = t_0$, then $y(t_1) = y(t_0)$.

4. Problem Formulation

In the interest of keeping notation simple we consider only nonlinear time-invariant circuits consisting of independent current sources and voltage controlled resistors and capacitors. These restrictions are mostly cosmetic, they allow the use of simple nodal analysis to formulate the circuit equations. All the results in this chapter can be applied to circuits containing inductors, voltage sources, and current-controlled components if a more general equation formulation method such as modified nodal analysis is used [sangiovanni81]. Initially we also only consider nonautonomous (or forced) circuits. That is, circuits with at least one nonconstant periodic or quasiperiodic input source. Within these constraints, two test problems are defined, one appropriate only for lumped devices, and one that allows linear distributed devices.

4.1. Lumped Test Problem

Let N be the number of nodes in a lumped circuit (excluding the reference node, ground). Consider two cases. For the periodic case, assume the input waveforms and the solution $\hat{v}$ belong to $P^N(T; \mathbb{R})$. For the quasiperiodic case with d fundamental frequencies λ_1, $\lambda_2, \ldots, \lambda_d$, assume both the input and the solution waveforms are members of $QP^N(\lambda_1, \lambda_2, \ldots, \lambda_d; \mathbb{R})$. Further assume that for both cases the solution is isolated and asymptotically stable and that all device constitutive equations are differentiable when written as a function of voltage. Now the circuit can be described by

$$f(v,t) = i(v(t)) + \dot{q}(v(t)) + u(t) = 0 \qquad (3.11)$$

where v is the vector of node voltage waveforms; u is the vector of source current waveforms; $i, q : \mathbb{R}^N \to \mathbb{R}^N$ are differentiable functions representing respectively the sum of the currents entering the nodes from *all* conductors, and the sum of the charges entering the nodes from *all* capacitors; f is the function that maps the node voltage waveforms into the sum of the currents entering each node; $t \in \mathbb{R}$ is time; and $0 \in \mathbb{R}^N$ is the zero vector.

4.2. Distributed Test Problem

Again consider two cases, the periodic case (as above) and the quasiperiodic case where the input and solution waveforms belong to $QP^N(\lambda_1, \lambda_2, \ldots, \lambda_d; \mathbb{R})$. Assume all other conditions for (3.11) hold except that the circuit may contain voltage controlled linear distributed devices. This circuit can be described by

$$f(v,t) = i(v(t)) + \dot{q}(v(t)) + \int_{-\infty}^{t} y(t-\tau)v(\tau)d\tau + u(t) = 0 \qquad (3.12)$$

where f is the function that maps the node voltage waveforms into the sum of the currents entering each node; $u \in QP^N(\lambda_1, \lambda_2, \ldots, \lambda_d; \mathbb{R})$

is the vector of source current waveforms; $i, q : \mathbb{R}^N \to \mathbb{R}^N$ are differentiable functions representing, respectively, the sum of the currents entering the nodes from the *nonlinear* conductors, and the sum of the charge entering the nodes from the *nonlinear* capacitors; y is the matrix-valued impulse response of the circuit with *all* the nonlinear devices removed[3], $t \in \mathbb{R}$ is time; and $0 \in \mathbb{R}^N$ is the zero vector.

Notice that in this test case, y represents all linear components and i and q represent only nonlinear components. This differs from the lumped test case where i represents all resistors (both linear and nonlinear) and q represents all capacitors.

5. Differential Equations

In general, differential equations have an infinite number of solutions and it is necessary to place constraints on the solutions until only one remains. If the constraints are all placed at the same point in time, the combination of the differential equation and the constraint equation is called an *initial-value problem* because the constraints are normally placed at the beginning of the interval of interest and the differential equation integrated with t increasing. It is also possible to put the constraints at the end of the interval of interest. Such a problem is referred to as a *final-value problem* and is treated identically to the initial-value problem except the differential equation is integrated with t decreasing. It is also possible to place the constraint at a point inside the interval of interest and break the problem into two independent problems, an initial- and a final-value problem.

It is not necessary to place all of the constraints at the same point in time. A differential equation combined with an algebraic equation that constrains the solution at two or more distinct points in time is referred to as a *boundary-value problem*.

---

[3]To remove a nonlinear device, simply replace its constitutive equation $y = f(x)$ with $y = 0$.

5.1. Initial-Value Problems

Consider the equation

$$\dot{x}(t) = f(x(t), t) \qquad (3.13)$$

where $x(t) \in \mathbb{R}^N$ is the state vector, and t is time. It is enough to specify the value of x at some point in time (ex., $x(0) = x_0$) and require that f be sufficiently smooth to assure that (3.13) has a unique solution.

Theorem 3.1: *Let the function $f(x,t)$ be continuous in t over the finite interval $[0,T]$ for all x and Lipschitz continuous in x, uniformly in t. That is,*

$$\| f(x,t) - f(y,t) \| \leq K \| x - y \|$$

for all $x,y \in \mathbb{R}^N$ and $t \in [0,T]$. K is known as the Lipschitz constant. Then the initial-value problem

$$\dot{x}(t) = f(x(t), t), \qquad x(0) = x_0$$

has a unique solution $x = x(t,x_0)$ over the interval $0 \leq t \leq T$. Furthermore, the solution is Lipschitz continuous in x_0, uniformly in t and satisfies

$$\| x(t,x_0) - x(t,y_0) \| \leq e^{Kt} \| x_0 - y_0 \|$$

for all $t \in [0,T]$ and $x_0, y_0 \in \mathbb{R}^N$.
□

The proof of this theorem is given in Stoer and Bulirsch [stoer80].

It is useful to know when a solution to the initial-value problem defined by (3.11) and an initial state $v(0) = v_0$ exists and is unique. Theorem 3.1 cannot be used directly because (3.11) is in the wrong form. However, it can be manipulated into the correct form by using the chain rule.

$$i(v(t)) + \frac{dq(v(t))}{dv(t)} \frac{dv(t)}{dt} + u(t) = 0$$

$$i(v(t)) + C(v(t))\dot{v}(t) + u(t) = \mathbf{0}$$

where $C(t) = \dfrac{dq(v(t))}{dv(t)}$, C being a time varying capacitance matrix. Assume $C(v(t))$ is nonsingular for all $v(t)$.

$$\dot{v}(t) = -C^{-1}(v(t))(i(v(t)) + u(t))$$

Thus from Theorem 3.1, the solution to the initial-value problem defined by (3.11) and the initial state $v(0) = v_0$ exists and is unique if i is Lipschitz continuous and if u is bounded and $C^{-1}(v(\cdot))$ is uniformly bounded with respect to v [white86].

5.2. Boundary-Value Problems

The problem of finding a solution to a system of ordinary differential equations is a boundary-value problem if that solution is required to satisfy subsidiary conditions at two or more distinct points in time. For example, finding the solution to

$$\dot{x}(t) = f(x(t), t) \tag{3.14 a}$$

over the interval $[0, T]$ is a two-point boundary-value problem if the solution is required to satisfy

$$g(x(0), x(T)) = \mathbf{0}. \tag{3.14 b}$$

Boundary-value problems are interesting when solving for steady-state solutions because the problem of finding the periodic or quasiperiodic solution to a differential equation can be posed as a boundary-value problem.

5.2.1. Existence and Uniqueness of Solutions

From Theorem 3.1, the initial-value problem

$$\dot{x}(t) = f(x(t), t), \qquad x(0) = x_0$$

is assured of having a unique solution if f is continuous in t and Lipschitz continuous in x uniformly in t. However, from the second part of this same theorem, it can be shown that these same conditions are insufficient to assure that the two-point boundary-value problem even has a solution. For example, consider

$$\dot{x}(t) = x(t) \qquad (3.15)$$

whose solution must satisfy

$$x(t) = e^t x(0). \qquad (3.16)$$

The boundary constraint

$$x(T) - e^T x(0) = 1 \qquad (3.17)$$

cannot be satisfied, as can be seen easily by using (3.16) to eliminate $x(T)$ from (3.17).

The existence and uniqueness theory for boundary-value problems is considerably more complicated and less thoroughly developed than for initial-value problems. If the solution is linearly constrained at only two points, which is the usual case, then some reasonably concise and powerful statements can be made. Consider a system of N first order differential equations

$$\dot{x}(t) = f(x(t), t) \qquad (3.18\,\mathrm{a})$$

subject to the most general linear two-point boundary conditions

$$Ax(0) + Bx(T) = c, \qquad (3.18\,\mathrm{b})$$

where $x(t) \in \mathbb{R}^N$, $f : \mathbb{R}^{N+1} \to \mathbb{R}^N$, $A, B \in \mathbb{R}^{N \times N}$, and $c \in \mathbb{R}^N$.

The study of the existence and uniqueness of solutions to the boundary-value problem (3.18) reduces to the study of the roots of a system of nonlinear equations by posing (3.18 a) as an initial-value problem. Consider

$$\dot{x}(t) = f(x(t), t) \qquad (3.19\,\mathrm{a})$$

$$x(0) = x_0 \qquad (3.19\,\mathrm{b})$$

and recall that the state-transition function $\phi(x_0, t_0, t_1)$ is the solution to this initial-value problem at time t_1 starting from the state x_0 at time t_0.

The solution to the boundary-value problem (3.18) is the solution to the initial-value problem (3.19) where the x_0 is chosen to satisfy the implicit nonlinear equation

$$Ax_0 + B\phi(x_0, 0, T) - c = 0. \tag{3.20}$$

In other words, if x_0 is a root of (3.20), then $x(t) = \phi(x_0, 0, t)$ solves (3.18).

Theorem 3.2: *Let the function $f(x,t)$ be continuous in t over the interval $[0,T]$ for all x and Lipschitz continuous in x, uniformly in t. Then the boundary-value problem (3.18) has as many solutions as there are distinct roots $x_0^{(j)}$ of Equation (3.20). These solutions are*

$$x^{(j)}(t) = \phi(x_0^{(j)}, 0, t),$$

the solutions of the initial-value problem (3.19) with initial state $x(0) = x_0^{(j)}$.
□

This theorem was developed by Keller and the proof is given in [keller68].

Reducing the problem of solving the boundary-value problem (3.18) to that of finding the roots of a system of nonlinear equations is interesting because it is the idea behind shooting methods. Shooting methods, presented in detail in Chapter 4, are the most commonly used methods for finding the solutions to boundary-value problems.

Keller, Stoer and Bulirsch present further theorems on the existence of solutions to boundary-value problems [keller68, stoer80].

5.3. Applications in Circuit Simulation

One approach to finding the steady-state solution of a circuit is to restrict the solution to be either periodic or quasiperiodic by using an appropriate boundary constraint. Thus, the problem of finding the steady state is converted into a boundary-value problem, to which the standard approaches, such as shooting methods and finite-difference methods, can be applied. In the following section, boundary constraints are formulated for periodic solutions. Boundary conditions for quasiperiodic solutions are presented in Chapter 7.

5.3.1. Periodic Boundary Constraint

A function x is periodic with period T if $x(t) = x(t + T)$ for *all* t. This is a difficult condition to apply in practice because it must be verified over all t. However, if x is the solution of a differential equation that satisfies the smoothness conditions of Theorem 3.1, then by uniqueness, if $x(t) = x(t + T)$ for some t, it is true for all t.

Theorem 3.3: *Consider the initial-value problem*

$$i(v(t)) + \dot{q}(v(t)) + u(t) = 0 \qquad (3.21\,a)$$

$$v(0) = v_0, \qquad (3.21\,b)$$

where $u(t)$ is T-periodic. If for all v_0 there exists a unique solution, and if there exists a solution v that satisfies

$$v(t) = v(t + T) \qquad (3.21\,c)$$

for some t, then v is T-periodic.

$\square$

This follows directly from the fact that $u(t + T) = u(t)$ for all t and the uniqueness assumption of the solution to the differential equation. Thus, to find the periodic steady-state solution of the lumped test problem, it is enough that the solution satisfy the two-point boundary constraint $v(t) = v(t + T)$ for *some* t and that (3.11) has a unique solution when formulated as an initial-value problem.

By Theorem 3.2, finding the periodic solution of a differential equation using the periodic boundary constraint (3.21 c) is equivalent to solving for the root of

$$\phi(v(0), 0, T) - v(0) = 0 \qquad (3.22)$$

where ϕ is the state-transition function of the differential equation. This equation takes the form of (3.20) with $A = -1_N$, $B = 1_N$, $c = 0$, and $x = v$, where 1_N is the $N \times N$ identity matrix.

5.3.2. Oscillators

A circuit is said to be *autonomous* if all its components and inputs do not vary with time. That is, the relationship between voltage and current for a resistor, voltage and charge for a capacitor, and flux and current for an inductor is time invariant and independent sources are constant valued. *Oscillators* are autonomous circuits that have nonconstant periodic solutions. The problem of finding the steady-state solution of an oscillator can be posed as a boundary-value problem with a free boundary. It is possible to adapt each of the methods that will be presented for two-point boundary-value problems to the oscillator problem.

Oscillators present two problems not previously faced. The first, as pointed out above, is that the period of the oscillation is unknown and must be determined. The second is that the time origin is arbitrary and thus if one solution exists, then an infinite continuum of solutions exists. In other words, if v is a solution, then so is any time shifted version of v (there is no input signal to fix the phase). If the oscillator is linear, the problem because even more difficult. For linear oscillators, the amplitude of the oscillation is not unique; if the circuit supports an undamped oscillation then the oscillation may be of any amplitude. Thus, not only is there a continuum of solutions parameterized on t, but also a continuum parameterized on the amplitude of the oscillation. A similar situation arises when a solution is a constant waveform. For this case, the additional continuum is parameterized in the period T.

Surprisingly, the existence of a continuum of solutions makes it difficult for many of the numerical methods used to compute steady-state to find any solution at all. We will return to this point at the end of the next chapter.

In general, one needs to worry about quasiperiodic solution to autonomous problems. Examples include driven oscillators where $\lambda_1, \ldots, \lambda_j$ are determined by autonomous oscillations and $\lambda_{j+1}, \ldots, \lambda_d$ are determined from input sources; and coupled oscillators, where $\lambda_1, \ldots, \lambda_d$ are all autonomous oscillations. The methods developed for periodic autonomous oscillators and quasiperiodic forced circuits are easily combined to handle these situations.

5.4. Numerical Solution of Initial-Value Problems

Lumped circuits are modeled using a system of nonlinear differential equations. Before the system can be solved it is necessary to specify the boundary conditions. Traditionally, circuit simulators treat only initial-value problems, so the initial state (here, the node voltages) are specified at $t = 0$ and the equations are integrated forward in time. Equation (3.11) can be written as an initial-value problem if the steady-state constraint on v is lifted and an initial state is specified.

$$i(v(t)) + \dot{q}(v(t)) + u(t) = \mathbf{0} \qquad\qquad v(0) = v_0 \qquad (3.23)$$

It is not possible in general to solve numerically systems of nonlinear differential equations directly. Instead, it is common to solve a discretized approximation to the differential equation.

5.4.1. Discretization

Discretization approximates a system of differential equations with a system of difference equations. In other words, the time interval of interest $[0,T]$ is divided into a finite number of possibly nonuniform subintervals with a monotonically increasing sequence of time-points $\{t_0, t_1, \cdots, t_S\}$ where $t_0 = 0$ and $t_S = T$. The subintervals are called time-steps and denoted by $h_s = t_s - t_{s-1}$. At t_s the solution of the discretized system v_s is an approximation to $v(t_s)$, the solution of the original differential equation (3.23). At each time-point, an algebraic system of equations must be solved, thus discretization converts a differential equation into a sequence of algebraic equations.

One commonly used approach to converting a system of differential equations into a sequence of systems of algebraic equations is to replace $\dot{q}$ in (3.23) with a finite-difference approximation [chua75] [gear71]. The simplest methods of this type are the one-step explicit and implicit Euler algorithms, defined by

$$\text{Explicit Euler: } \dot{q}_s = \frac{q_{s+1} - q_s}{h_s} \tag{3.24}$$

$$\text{Implicit Euler: } \dot{q}_{s+1} = \frac{q_{s+1} - q_s}{h_{s+1}} \tag{3.25}$$

Of these two methods, implicit (or backward) Euler has the desirable property of being stiffly stable, which means that it is well behaved even if the differential equations being solved have widely separated time constants. This is not true for explicit (or forward) Euler [white86] [gear71].

It is natural to apply methods like explicit and implicit Euler to initial-value problems because, when computing the solution at t_s, it is only necessary to know the solution at previous values of time. The integration is conveniently carried out by starting at t_0 and progressing forward in time toward t_S.

6. Numerical Solution of Nonlinear Equations

The algebraic equations generated by discretization are nonlinear and in general are not solvable explicitly. Instead, the equations are solved via some kind of iterative improvement technique, where a first guess is made at the solution, how well the guess satisfies the equations is measured, then some adjustment is made to the guess based on that measure. The process is repeated until the solution satisfies the equations "well enough".

One approach to iterative improvement is to correct a guess at the solution by solving a system generated by linearizing the nonlinear equation about the guessed solution. In this way, solving the nonlinear problem is converted to solving a sequence of linear problems constructed such that, if an accumulation point of the iteration exists, its solution is the solution of the nonlinear problem. Such methods are referred to as Newton-Raphson methods.

To derive the Newton-Raphson iteration equations precisely, consider the implicit nonlinear system of equations

$$f(\hat{x}) = 0 \tag{3.26}$$

where $\hat{x} \in \mathbb{R}^N$ and $f : \mathbb{R}^N \rightarrow \mathbb{R}^N$ is Lipschitz continuously differentiable. By Taylor's theorem, an initial guess $x^{(0)}$ must satisfy

$$f(\hat{x}) = f(x^{(0)}) + \frac{df(x^{(0)})}{dx}(\hat{x} - x^{(0)}) + O((\hat{x} - x^{(0)})^2) \tag{3.27}$$

where $\dfrac{df(x)}{dx}$ is the Frechet derivative of f with respect to x [ortega70]. The Jacobian $J_f(x)$ is a representation of the Frechet derivative, and if it exists it takes the form

$$\frac{df(x)}{dx} = J_f(x) = \left[\frac{\partial f_m(x)}{\partial x_n} \right], \qquad m, n = 1, 2, \ldots, N.$$

$O(\cdot)$ is a function that represents the higher order terms and is such that $\lim\limits_{\alpha \to 0} \dfrac{\| O(\alpha) \|}{\alpha}$ is bounded. Equation (3.27) suggests that if $(\hat{x} - x^{(0)})$ is small then the root of

$$f_L^{(0)}(x^{(1)}) = f(x^{(0)}) + J_f(x^{(0)})(x^{(1)} - x^{(0)}),$$

where $f_L^{(0)}(x)$ is the linearized approximation to $f(x)$ (it is the hyperplane that is tangent to $f(x)$ at $x^{(0)}$), will be close to $\hat{x}$. An improved approximation to $\hat{x}$ is now found by solving for the root of the linearized approximation to $f(x)$. That is, finding the value of x, denoted by $x^{(1)}$, that satisfies $f_L^{(0)}(x) = 0$. This $x^{(1)}$ is given by

$$x^{(1)} = x^{(0)} - J_f^{-1}(x^{(0)})f(x^{(0)}).$$

If $f_L^{(0)}$ is a good approximation to f near $\hat{x}$, $x^{(1)}$ will be closer to $\hat{x}$ than was $x^{(0)}$. This procedure is repeated by replacing the initial guess $x^{(0)}$ with $x^{(1)}$. The iteration, as shown in Figure 3.2, is repeated until some convergence criteria is satisfied.

$$x^{(j+1)} = x^{(j)} - J_f^{-1}(x^{(j)})f(x^{(j)}) \tag{3.28}$$

The sequence generated by (3.27) converges to $\hat{x}$ if f is continuously differentiable, $J_f(\hat{x})$ is nonsingular, and $x^{(0)}$ is sufficiently close to $\hat{x}$ [dahlquist74]. Furthermore, if $J_f(x)$ is Lipschitz continuous, the asymptotic rate of convergence will be at least quadratic. More

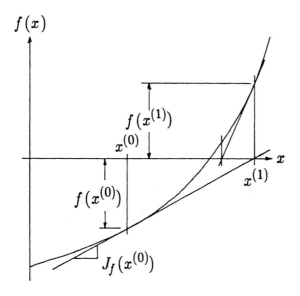

Figure 3.2: Newton-Raphson illustrated.

explicitly, let $\varepsilon^{(j)} = \| x^{(j)} - \hat{x} \|$ be the error of the jth iterate. If there exist constants p and $\alpha \neq 0$ such that

$$\lim_{j \to \infty} \frac{\varepsilon^{(j+1)}}{(\varepsilon^{(j)})^p} = \alpha,$$

then p is called the rate (or order) of convergence. For Newton-Raphson, when $J_f(x)$ is Lipschitz, $p \geq 2$. In general, there is no way to assure that the initial guess $x^{(0)}$ is sufficiently close to the solution $\hat{x}$, so convergence can be elusive.

6.1. Convergence Criteria

As a practical matter, it is necessary to terminate the Newton-Raphson iteration after a finite number of iterations. It is natural to stop the iteration when $f(x^{(j)})$ is negligible. More formally, the iteration stops when

$$\| f(x^{(j)}) \| < \varepsilon_f \qquad (3.29)$$

where ε_f is some small positive number. To see how this convergence criterium affects the error in the solution, assume that the true solution is $\hat{x}$ and that $\| f(x^{(j)}) \| \le \varepsilon_f$. By expanding f about $x^{(j)}$, it is easy to show that to first order,

$$J_f(x^{(j)})[\hat{x} - x^{(j)}] \approx -f(x^{(j)}). \qquad (3.30)$$

Thus the quantity $\Delta x^{(j)} = -J_f^{-1}(x^{(j)})f(x^{(j)})$ is a first order estimate of the error in the solution. Notice that this quantity has been previously computed in (3.28). Thus, at very little added cost, the following additional convergence criterium can be added to directly control the error in the solution

$$\| \Delta x^{(j)} \| < \varepsilon_x \qquad (3.31)$$

It is best to use both (3.29) and (3.31) as convergence criteria for Newton-Raphson because if only (3.29) is used it is possible to have (3.26) nearly satisfied but still have a large error in the solution (this would occur if J_f was ill-conditioned). If only (3.31) is used, it is possible to terminate the iteration prematurely when (3.26) is far from being satisfied because progress toward the solution on one step is slow (and therefore Δx is small). This last situation is referred to as *false convergence*.

6.2. Simplified Newton-Raphson Methods

Newton-Raphson requires that the Jacobian be constructed and factored on each iteration; an expensive series of operations that, under certain circumstances, can be avoided. Several simplified Newton-Raphson methods have been developed over the years that reduce this expense. We are interested here the one in which the Jacobian is formed and factored only for the first iterate.

$$x^{(j+1)} = x^{(j)} - J_f^{-1}(x^{(0)})f(x^{(j)}) \qquad (3.32)$$

If the function f is near-linear then the changes in the Jacobian from iteration to iteration are small and the first Jacobian closely approximates the true Jacobian on subsequent steps. In this case, this new method converges to the correct solution. The Jacobian is only used to generate new iterates, and is not used when confirming convergence, so errors resulting from approximations in the Jacobian only affect the rate and region of convergence, not the accuracy of the final solution. This simplified Newton-Raphson method can often be considerably faster than standard Newton-Raphson with large near linear systems of equations, even though it usually requires a greater number of iterations, because each iteration is less expensive. However, the region of convergence is often smaller than with standard Newton-Raphson.

To increase the region of convergence over the above simplified Newton-Raphson method, reevaluate the Jacobian every k iterations rather than use the original Jacobian until convergence. This method, referred to as Samanskii's method, may be considered a k step method, where each iterate is composed of one Newton-Raphson step and $k - 1$ simplified Newton-Raphson steps. Traub and Samanskii [ortega70] showed that sequence consisting of every k^{th} iterate converges with order $k + 1$. Thus, if $k = 2$, that is, if the Jacobian is updated on every other iterate, then cubic convergence is achieved. As expected, this rate of convergence is inferior to that of conventional Newton-Raphson, where the sequence of consisting of every other iterate converges quarticly.

The advantages of the simplified Newton-Raphson iterates is combined with the large region of convergence of conventional Newton-Raphson by monitoring the sequence $\{ \| f(x^{(j)}) \| \}$ and using simplified Newton-Raphson iterates as long as a sufficient reduction in this norm is achieved. If a simplified iteration results in an insufficient reduction, the iterate and the old Jacobian should be discarded and a full Newton-Raphson step taken.

6.3. Continuation Methods

In order to assure that Newton-Raphson converges, it is necessary to supply an initial guess that is sufficiently close to the solution. Since, in general, the solution is not known in advance, it is often difficult to supply such a guess. *Continuation methods* provide a way for obtaining starting points that are sufficiently close to assure convergence.

Usually, the problem depends in a natural way on some parameter p, such that when the parameter is set equal to some specific value, say 1, the particular system for which the solution is desired results, while for $p = 0$ the system has a known solution x_0. Thus,

$$f(x,p) = 0 \tag{3.33}$$

where $f(x,1) = f(x)$ and where $f(x_0,0) = 0$. Assume that in (3.33), x can be written as a function of p, i.e., $f(x(p),p) = 0$. Then the solution $x(p)$ can be found for an increasing sequence of values of p, $0 = p_0 < p_1 < p_2 < \cdots < p_S = 1$. If $x(p)$ is a continuous function of p, then it is always possible to choose p_s close enough to p_{s-1} so that if $x(p_{s-1})$ is used as a starting point, it is sufficiently close to $x(p_s)$ to assure convergence. The fundamental idea in continuation methods is to generate a finite sequence of problems, the solution to the first of which is known, and such that the solutions of each problem is close enough to the solution of the next to be within the region of convergence for Newton-Raphson on the next. The step size $p_s - p_{s-1}$ should be adjusted on each step to minimize the total number of Newton iterations rather than the number of steps.

Linear extrapolation is normally used with continuation methods to reduce the number of steps required. After the first step, a simple form of linear extrapolation can be used based on the solution at the previous two steps.

$$x^{(0)}(p_s) = x^{(0)}(p_{s-1}) + \frac{p_s - p_{s-1}}{p_{s-1} - p_{s-2}}[x(p_{s-1}) - x(p_{s-2})]$$

A more sophisticated linear extrapolation is performed by using the derivative $\partial x / \partial p$.

$$x^{(0)}(p_s) = x^{(0)}(p_{s-1}) + (p_s - p_{s-1})\frac{\partial x(p_{s-1})}{\partial p}$$

where

$$\frac{\partial x(p_{s-1})}{\partial p} = -\left[\frac{\partial f(x(p_{s-1}),p_{s-1})}{\partial x}\right]^{-1}\frac{\partial f(x(p_{s-1}),p_{s-1})}{\partial p},$$

which can be derived by using the implicit function theorem [rudin76]. This approach is usually preferred if the derivatives are readily available.

When computing the DC operating point of a circuit, it is very common to use as a continuation parameter the fraction of the DC source voltages and currents applied to the circuit. In almost all circuits simulators, all devices except sources pass no current when all terminal potentials are zero. Thus, when all sources are turned off, circuits are guaranteed to have a solution with zero potential at every node and zero current through every branch. From this known solution, the source levels can be slowly increased, while solving the circuit at each step, until the desired source levels are attained. Continuation implemented in this manner is generally referred to as *source stepping*.

Continuation can fail if $x(p)$ is not a continuous function on the interval [0,1]. By the implicit function theorem, if $\frac{\partial f(x(p),p)}{\partial x}$ exists and is continuous over a neighborhood of (x_0,p_0) and is nonsingular at (x_0,p_0), then there exists a neighborhood of (x_0,p_0) for which x is a continuous function of p and $f(x(p),p) = 0$. Thus, continuation methods will be successful if $\frac{\partial f(x(p),p)}{\partial x}$ is continuously differentiable and is nonsingular at $x(p)$ for all $p \in [0,1]$ It is not always possible to assure this in practice.

Consider the circuit in Figure 3.3. The supply potential versus supply current is plotted in Figure 3.4. Supply potential is not a function of input current because there is not a unique potential for each current. Indeed, source stepping would fail on this circuit when the supply reached $p_0 = 1.6mA$ because the potential would have to jump discontinuously from 11V to 28V. At this point, which is called a *limit point*, the Jacobian $J_f(x(p)) = \dfrac{\partial f(x(p),p)}{\partial x}$ is singular.

Continuation methods fail at limit points for two reasons. First, the Jacobian $J_f(x(p))$ is singular at limit points, which causes Newton-Raphson to fail. If this were the only problem, limit points would not be a serious difficulty because singularities are isolated points and it would be possible to step beyond them. However, this is prevented by the second reason. Monotonically increasing the parameter p results in a discontinuous jump in the solution at a limit point. Such problems can still be solved, but require the more sophisticated arc-length continuation methods described in Appendix C.

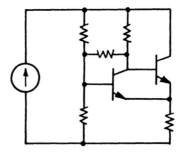

Figure 3.3: A circuit that causes trouble for simple source-stepping algorithms.

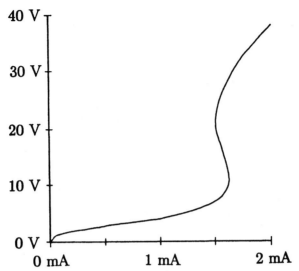

Figure 3.4: Supply potential versus supply current for the circuit in Figure 3.3.

Chapter 4
Time-Domain Methods

A very important approach to finding the steady-state response of a system of differential equations is to formulate a boundary-value problem whose solution is the desired steady-state response. Methods for formulating boundary-value problems for periodic solutions was discussed in Chapter 3 and methods for quasiperiodic solutions will be covered in Chapter 7. This chapter presents two important ways of solving these boundary-value problems.

1. Finite-Difference Methods

Like most approaches to solving differential equations numerically, the finite-difference methods approximate the original system with a set of difference equations. Unlike finite-difference methods for transient analysis methods, finite-difference methods for boundary-value problems attempt to find the solution at every time-point simultaneously. To find a T-periodic solution using a finite-difference method, a mesh $t_0 < t_1 < t_2 < \cdots < t_S$ is chosen where $t_0 = 0$ and $t_S = T$. Assume for simplicity that the time-points are equally spaced. A finite sequence $\{v_s\}$ is computed as an approximation to $v(t)$ on the mesh, where $v_s \approx v(t_s)$. The difference equations are formed by using a discrete-time approximation to the time derivative and the convolution integral. For example, consider the slightly modified version of (3.12)

$$f(v,t) = i(v) + \dot{q}(v) + \int_{-\infty}^{t} y(t-\tau)v(\tau)d\tau + u(t) = \mathbf{0}. \qquad (4.1)$$

All symbols have their previous definitions except that i is a function representing the sum of the current entering the nodes from *all* conductors, q represents the sum of the charge entering the nodes from *all* capacitors; and y is the matrix-valued impulse response of the circuits with all *lumped* elements removed.

There are a large number of possible discrete approximations to $\dot{q}$ that can be used [gear71]. Implicit Euler [white86], the simplest approximation that is suitable for circuit simulation, employs linear interpolation between mesh points and is given by

$$\dot{q}_s = \frac{1}{h_s}(q_s - q_{s-1}) \qquad (4.2)$$

where $h_s = t_s - t_{s-1}$ is the time-step. A discrete approximation to the convolution integral would be given by

$$w_s = \sum_{r=0}^{s} y_{s-r} v_r \qquad (4.3)$$

where w_s represents the current entering the nodes from distributed devices at time t_s and y_s is the discrete approximation to the impulse response y. Computation of y is done by evaluating a phasor representation of the distributed device, $Y(j\omega)$, over a range of frequencies. The data is then windowed [harris78], inverse transformed [djordjevic86, schutt-aine88], and periodically extended. If the mess of time-points were not equally spaced, the computation of w_s would also involve applying an interpolation function to the solution between the mesh points. The computation could be refined by employing an integration method such as Simpson's rule to compute w_s [hall76, stoer80].

Discretizing (4.1) using the approximations of (4.2) and (4.3) yields

$$i(v_s) + \frac{1}{h}(q(v_s) - q(v_{s-1})) + \sum_{r=0}^{S} y_{s-r} v_r + u_s = 0 \qquad (4.4)$$

where $s = 1, \ldots, S$ and h is the time-step. As v is T-periodic, $v_0 = v_S$, which results in the following system of nonlinear algebraic equations

$$i(v_1) + \frac{q(v_1) - q(v_S)}{h} + \sum_{r=0}^{S} y_{1-r} v_r + u_1 = 0$$

$$i(v_2) + \frac{q(v_2) - q(v_1)}{h} + \sum_{r=0}^{S} y_{2-r} v_r + u_2 = 0$$

$$\vdots \qquad \qquad \vdots \qquad \qquad \vdots \qquad \vdots \quad \vdots$$

$$i(v_S) + \frac{q(v_S) - q(v_{S-1})}{h} + \sum_{r=0}^{S} y_{S-r} v_r + u_S = 0 .$$

The system is solved using the Newton-Raphson algorithm.

The finite-difference methods are elegant and simple. When applied to two-point boundary-value problems with linear boundary constraints, such as the periodicity constraint, they have the characteristic that each iterate generated by the Newton-Raphson algorithm satisfies the boundary condition, though it may not satisfy the difference equation. This contrasts with shooting methods, for which each iterate satisfies the difference equation but not the boundary conditions.

Finite-difference methods can generate large systems of equations, especially if either the number of unknown waveforms or the number of time-points is large. The systems are sparse, and hence, not overly expensive to solve. The sparsity is increased further if there are no distributed devices present in the circuit. In this case the equations take on an interesting structure. To see this, consider implicit Euler applied to the lumped test problem (3.11).

$$f(v,t) = \dot{q}(v(t)) + i(v(t)) + u(t) = 0 \qquad (4.5)$$

to generate

$$\frac{1}{h_{s+1}}(q(v_{s+1}) - q(v_s)) + i(v_{s+1}) + u_{s+1} = 0 \qquad (4.6)$$

where $s = 0, \ldots, S-1$ and the time-points are no longer assumed to be equally spaced. As v is T-periodic, setting $v_0 = v_S$ results in the following system of nonlinear algebraic equations

$$\frac{q(v_1) - q(v_S)}{h_1} + i(v_1) + u_1 = 0$$

$$\frac{q(v_2) - q(v_1)}{h_2} + i(v_2) + u_2 = 0$$

$$\vdots \qquad\qquad \vdots \quad \vdots \quad \vdots$$

$$\frac{q(v_S) - q(v_{S-1})}{h_S} + i(v_S) + u_S = 0 \,.$$

As usual, the system is solved using Newton-Raphson. The solution waveform $\{v_s\}$ is found by taking this iteration to convergence.

It is possible to reduce the time required for the finite-difference methods by choosing carefully the time-steps to achieve a desired accuracy, clustering them in troublesome spots to reduce error while spreading them out in quiescent areas to reduce computer resource usage. Generally this is done by starting with a small number of time-points, and adding more in areas exhibiting excessive error in later iterations. If the solution is expected to be smooth, it is possible to use high-order integration methods to achieve low truncation error using widely separated time-points. Lastly, it is possible to use the Šamanskii's method [ortega70] to reduce execution time further. The possible efficiency gain for Šamanskii's method is considerable if a high order integration method is used (and so the bandwidth of the Jacobian is large and therefore expensive to factor) or if the Jacobian is expensive to construct.

$$
\begin{bmatrix}
\dfrac{c(v_1^{(j)})}{h_1} + g(v_1^{(j)}) & & & & \dfrac{-c(v_S^{(j)})}{h_1} \\[2ex]
\dfrac{-c(v_1^{(j)})}{h_2} & \dfrac{c(v_2^{(j)})}{h_2} + g(v_2^{(j)}) & & & \\[2ex]
& & \ddots & & \\[2ex]
& & & \dfrac{-c(v_{S-1}^{(j)})}{h_S} & \dfrac{c(v_S^{(j)})}{h_S} + g(v_S^{(j)})
\end{bmatrix}
$$

$$
\begin{bmatrix}
v_1^{(j+1)} - v_1^{(j)} \\
v_2^{(j+1)} - v_2^{(j)} \\
\vdots \\
v_S^{(j+1)} - v_S^{(j)}
\end{bmatrix}
= -
\begin{bmatrix}
\dfrac{q(v_1^{(j)}) - q(v_S^{(j)})}{h_1} + i(v_1^{(j)}) + u_1 \\[2ex]
\dfrac{q(v_2^{(j)}) - q(v_1^{(j)})}{h_2} + i(v_2^{(j)}) + u_2 \\[2ex]
\vdots \\[2ex]
\dfrac{q(v_S^{(j)}) - q(v_{S-1}^{(j)})}{h_S} + i(v_S^{(j)}) + u_S
\end{bmatrix}
$$

$$(4.7)$$

Finite-difference methods are well known in the mathematics community, but have received little attention from the circuit simulation community. Because of the large number of equations and unknowns, finite-difference methods require a large amount of memory, which often constrains the size of the circuit that can be simulated. Finite-difference methods also tend to have more Newton-Raphson convergence problems than the other popular method for solving boundary-value problems, shooting methods (presented next). The difference results from the shooting method ability to hide nonlinear behavior from the Newton-Raphson algorithm in some common situations.

One feature of the finite-difference methods is not shared by the shooting methods. Once a solution has been found by a finite-difference method, the resulting Jacobian represents a linearization of the circuit over the simulation interval. Thus, if finite-difference methods are used to find a periodic steady-state solution, the Jacobian is the linearization of the circuit about its periodically time-varying operating point. Just as the Jacobian that results from a DC analysis is the linearization of the circuit about its quiescent operating point that is used in small signal calculations such as the AC and noise analyses, the finite-difference method Jacobian can be used in small signal applications as well [sugawara89]. Since it is represents a time-varying linear circuit, it exhibits frequency conversion.

2. Shooting Methods

In general, shooting methods for solving boundary-value problems can be thought of as iterative methods applied to solving the nonlinear algebraic system generated by replacing the differential equation relation with the state-transition function. For example, consider the two-point boundary-value problem with linear constraints,

$$\dot{v}(t) = f(v(t), t) \tag{4.8}$$

subject to

$$Av(0) + Bv(T) = c,$$

By Theorem 3.2, solving this boundary-value problem is equivalent to solving the nonlinear algebraic problem

$$Av(0) + B\phi(v(0), 0, T) - c = 0. \tag{4.9}$$

for $v(0)$, where ϕ is the state-transition function for (4.8). As (4.9) is a nonlinear algebraic problem, a variety of standard methods, like fixed-point iteration or Newton-Raphson, can be used to compute $v(0)$. The form of (4.9) is more general than required. For simplicity, assume $A = -D$, $B = \mathbf{1}$, and $c = 0$. With these assumptions

$$\phi(v(0), 0, T) - Dv(0) = 0 \tag{4.10}$$

becomes our generic two-point boundary-value problem. Further assume that D is nonsingular. As shown in Chapter 3, the periodic solution of a differential equation can be found by solving (3.22), or

$$\phi(v(0), 0, T) - v(0) = 0. \tag{4.11}$$

This problem is mapped into (4.10) by setting $D = I_N$, the $N \times N$ identity matrix.

When looking for a periodic solution to the differential equation for the lumped test problem, it is possible to cast the obvious approach of integrating (3.11) until the transient decays to an acceptably small level, as a shooting method. A simple fixed-point iteration to solve (4.11) is

$$v^{(k+1)}(0) = \phi(v^{(k)}(0), 0, T), \tag{4.12}$$

which is equivalent to integrating the original differential equation from $v(0) = v_0$ until all transients decay.

There are two well known ways to accelerate the convergence of this iteration to the desired solution, extrapolation methods and Newton methods. In extrapolation methods, several fixed-point iterations, as in (4.12), are computed and progress towards steady state is monitored. The sequence constructed from the fixed-point iterates is used to accelerate the march towards steady state. Use of the extrapolation shooting methods in circuit simulation was championed by Skelboe [skelboe80].

The Newton methods apply the Newton-Raphson algorithm to (4.11), which involves computing the derivative of the final state with respect to changes in the initial state. The combination of Newton-Raphson and the shooting method was first used to find the periodic steady-state response of circuits by Aprille and Trick [aprille72b] [aprille72a].

Both of these approaches apply as well to the generic two-point boundary-value problem (4.10) as to the periodic problem.

2.1. Shooting by Extrapolation

Formulate (4.10) as a fixed-point iteration and construct a sequence $\{v_s\}$ by letting $v_0 = v(0)$ and using

$$v_{s+1} = D^{-1}\phi(v_s, 0, T). \qquad (4.13)$$

Convergence of this sequence can be accelerated by extrapolation. Though several extrapolation techniques can be used [skelboe80] [smith87], the one based on minimum polynomials is presented here. It is normally the most efficient because it requires the fewest periods to perform an extrapolation.

The extrapolation is made by assuming that the sequence $\{v_s\}$ is generated by a linear finite dimensional system of the form

$$v_{s+1} = Av_s + b \qquad (4.14)$$

where $A \in \mathbb{R}^{N \times N}$ and $b \in \mathbb{R}^N$. By just observing $\{v_s\}$ and without explicitly calculating A and b, the fixed-point v^* of this linear model can be found. If $\phi(\cdot, 0, T)$ is linear, then (4.14) models $\phi(\cdot, 0, T)$ exactly, $v^* = \hat{v}$, and the solution is found in one iteration. However if $\phi(\cdot, 0, T)$ is nonlinear, then more than one extrapolation iteration is usually needed. Expand $\phi(\cdot, 0, T)$ about its fixed-point $\hat{v}$:

$$\phi(\hat{v} + \delta v, 0, T) = \phi(\hat{v}, 0, T) + J_\phi(\hat{v}, 0, T)\delta v + O(\delta v^2)$$

$$= \phi(\hat{v}, 0, T) - J_\phi(\hat{v}, 0, T)\hat{v} +$$

$$J_\phi(\hat{v}, 0, T)(\hat{v} + \delta v) + O(\delta v^2).$$

Let $b = \phi(\hat{v}, 0, T) - J_\phi(\hat{v}, 0, T)\hat{v}$ and $A = J_\phi(\hat{v}, 0, T)$, then

$$\phi(\hat{v} + \delta v, 0, T) = b + A(\hat{v} + \delta v) + O(\delta v^2)$$

If δv is large, then $O(\delta v^2)$ is large and the extrapolated value v^* is different from $\hat{v}$. Presumably though, v^* is closer to $\hat{v}$ than was the initial guess v_0 and the extrapolation process can be repeated until it converges to $\hat{v}$. Convergence is achieved if v_0 is close enough to $\hat{v}$ and if $J_\phi(\hat{v}, 0, T)$ is nonsingular, and it is quadratic if $\phi(\cdot, 0, T)$ is sufficiently smooth [skelboe82].

Now the extrapolation algorithm is presented. Consider the sequence $\{\delta v_s\}$ generated by (4.14) where $\delta v_s = v_s - v^*$. It is easy to show that $\delta v_s = A\delta v_s$. For this sequence to converge, the spectral radius of A must be less than one, but if A has any eigenvalues with magnitude close to one, then convergence will be slow. The eigen-values with magnitude close to one correspond to large time constants in the circuit. Assume that there are p such eigenvalues, then the δv_s vectors will line up along the p eigenvectors associated with the slow eigenvalues asymptotically as s increases. It is possible to solve for v^* by solving a p order system. In particular, if the δv_s vectors line up along p eigenvectors, then there exists a set of $p + 1$ nonzero coefficients $\{c_s\}$ $s = 0, 1, \ldots, p$ such that

$$\sum_{s=0}^{p} c_s \delta v_s = 0. \tag{4.15}$$

If the coefficients c_s are found, then it is possible to compute the fixed-point v^* for (4.14) from (4.15) and the fact that $\delta v_s = v_s - v^*$.

$$v^* = \frac{\sum_{s=0}^{p} c_s v_s}{\sum_{s=0}^{p} c_s} \tag{4.16}$$

In order to find the coefficients c_s, define $\Delta v_s = v_{s+1} - v_s$ and recall that $v_{s+1} = A v_s + b$ and $v^* = A v^* + b$. Then

$$\sum_{s=0}^{p} c_s \Delta v_s = \sum_{s=0}^{p} c_s (A v_s + b - v_s),$$

$$\sum_{s=0}^{p} c_s \Delta v_s = \sum_{s=0}^{p} c_s (A v_s + v^* - A v^* - v_s),$$

$$\sum_{s=0}^{p} c_s \Delta v_s = \sum_{s=0}^{p} c_s (A - 1)(v_s - v^*),$$

$$\sum_{s=0}^{p} c_s \Delta v_s = (A - 1) \sum_{s=0}^{p} c_s (v_s - v^*) = 0.$$

$$\sum_{s=0}^{p} c_s \Delta v_s = \mathbf{0}. \qquad (4.17)$$

Thus, the same set of coefficients $\{c_s\}$ can be used with both $\{\delta_s\}$ and $\{\Delta_s\}$ to form linear combinations that sum to zero. This allows (4.17) to be used to determine the coefficients and (4.16) to be used to find v^*. To solve (4.17) for the coefficients $\{c_s\}$, let $V = [\Delta v_0, \Delta v_1, \ldots, \Delta v_{p-1}]$, $c = [c_0, c_1, \ldots, c_{p-1}]^T$ and $c_p = -1$. Then (4.17) becomes

$$Vc = \Delta v_p. \qquad (4.18)$$

If $p < N$, the problem is overdetermined and so cannot be solved using LU factorization. Even if $p = N$, one of more time constants may be short compared to the shooting interval, and so the corresponding eigenvalues will be small, resulting in V being ill-conditioned and making LU factorization risky. Thus (4.18) is treated as a least squares problem and is solved using QR factorization [dahlquist74]. In other words, the coefficient vector c is chosen by the QR algorithm to minimize ε, where

$$\varepsilon = \| Vc - \Delta v_p \|_2^2 \qquad (4.19)$$

Since A and its eigenvalues are unknown, there is no way to compute p explicitly. Instead, its value is estimated by monitoring ε as the number of periods in the computed response is increased. The value of p is taken as the smaller of either N or the number of periods used for the computation of c when ε drops below some small threshold. Once p has been determined, the extrapolation should be performed using (4.16) and (4.18).

The calculation of v^* represents one iterate of the extrapolation process. The iteration is continued until the sequence of v^*'s converges. Each iterate requires simulating the circuit for at least p iterations. The value of p is roughly equal to the number of independent slowly decaying states in the circuit, so if there are few, the extrapolation version of the shooting method should be efficient. In order to eliminate any effect of the short lived time constants, it is a good idea to discard the first period of the circuit response waveform in each extrapolation iterate.

2.2. Shooting with Newton-Raphson

When applying Newton-Raphson directly to (4.11), not only is it necessary to compute the response of the circuit over one period, the sensitivity of the final state with respect to changes in the initial state v_0 must also be computed. The sensitivity is used to determine how to correct the initial state once the difference between the achieved and the desired final state is found. Perhaps conceptually the easiest way to determine the sensitivity of the final state to the initial state is to use finite differences. In this case, the circuit equations are solved N times; each time perturbing slightly a different entry in the v_0 vector. The approach taken when using Newton method is similar except that the circuit is linearized about the v_0 solution trajectory and the resulting linear time-varying system is solved for its zero-input response with N different initial conditions. The N initial conditions are normally taken to be the N unit vectors that span $\mathbb{R}^N$ [aprille72b].

Applying Newton-Raphson to

$$\phi(\hat{v}_0, 0, T) - D\hat{v}_0 = 0 \tag{4.20}$$

results in the iteration

$$v_0^{(j+1)} = v_0^{(j)} - [J_\phi(v_0^{(j)}, 0, T) - D]^{-1}[\phi(v_0^{(j)}, 0, T) - Dv_0^{(j)}] \tag{4.21}$$

where j is the iteration number and

$$[J_\phi(v_0, 0, T) - D] = \frac{d}{dv_0}(\phi(v_0, 0, T) - Dv_0) = \frac{dv_S}{dv_0} - D. \tag{4.22}$$

There are two important pieces to the computation of the Newton iteration given in (4.21): factoring the matrix $[J_\phi(v_0, 0, T) - D]$, which is a full matrix in general, and evaluating the state-transition function $\phi(v_0, 0, T)$ and its Frechet derivative $J_\phi(v_0, 0, T)$. The state-transition function is computed by integrating (3.11) numerically over the shooting interval. The derivative of the state-transition function, referred to as the sensitivity matrix, is computed simultaneously because there are several quantities that are common to both computations, as explained below.

At this point it is important to realize that we are not solving the original circuit equation (3.11), but rather a discretized approximation. This distinction is important to achieving quadratic convergence in the Newton-Raphson iteration [skelboe82]. The notation needed to perform the derivation for the general backward difference discretization is complex enough to obscure the basic concepts behind the derivation, so for simplicity and clarity, (3.11) is discretized using implicit Euler (3.25), though any discretization formula could have been used instead.

$$f(v_s) = \frac{1}{h_s}\left[q(v_s) - q(v_{s-1})\right] + i(v_s) + u_s = 0 \qquad (4.23)$$

where $s = 1, 2, \ldots, S$, h_s is the time-step, $t_0 = 0$, $t_s = h_s + t_{s-1}$, and $t_S = T$.

$\phi(\cdot, 0, T)$ is found by evaluating (4.23) recursively, at each step solving the implicit set of nonlinear equations using Newton-Raphson.

$$J_f(v_s^{(j-1)})[v_s^{(j)} - v_s^{(j-1)}] = -f(v_s^{(j-1)})$$

$$\left[\frac{1}{h_s}\frac{dq(v_s^{(j-1)})}{dv_s} + \frac{di(v_s^{(j-1)})}{dv_s}\right]v_s^{(j)} - v_s^{(j-1)}$$

$$= -\frac{1}{h_s}\left[q(v_s^{(j-1)}) - q(v_{s-1})\right] - i(v_s^{(j-1)}) - u_s$$

Let $di(v)/dv = g(v)$ and $dq(v)/dv = c(v)$.

$$\left[\frac{c(v_s^{(j-1)})}{h_s} + g(v_s^{(j-1)})\right][v_s^{(j)} - v_s^{(j-1)}] \qquad (4.24)$$

$$= -\frac{1}{h_s}\left[q(v_s^{(j-1)}) - q(v_{s-1})\right] - i(v_s^{(j-1)}) - u_s$$

The sensitivity $\dfrac{dv_S}{dv_0}$ is taken to be the final value of the $\dfrac{dv_s}{dv_0}$ trajectory, which is found by differentiating both sides of (4.23) with respect to v_0.

$$\frac{1}{h_s}\frac{d}{dv_0}\left[q(v_s) - q(v_{s-1})\right] + \frac{d}{dv_0}i(v_s) = 0 \qquad (4.25)$$

Applying the chain rule

$$\frac{1}{h_s}\left[\frac{dq(v_s)}{dv_s}\frac{dv_s}{dv_0} - \frac{dq(v_{s-1})}{dv_{s-1}}\frac{dv_{s-1}}{dv_0}\right] + \frac{di(v_s)}{dv_s}\frac{dv_s}{dv_0} = 0.$$

Let $di(v)/dv = g(v)$ and $dq(v)/dv = c(v)$.

$$\left[\frac{c(v_s)}{h_s} + g(v_s)\right]\frac{dv_s}{dv_0} - \frac{c(v_{s-1})}{h_s}\frac{dv_{s-1}}{dv_0} = 0$$

$$\frac{dv_s}{dv_0} = J_f^{-1}(v_s)\frac{c(v_{s-1})}{h_s}\frac{dv_{s-1}}{dv_0} \qquad (4.26)$$

where $J_f(v_s) = \dfrac{c(v_s)}{h_s} + g(v_s)$.

The Jacobian $J_\phi(v_0, 0, T) = \dfrac{dv_s}{dv_0}$ is computed by repeated application of (4.26) starting from the initial condition $\dfrac{dv_0}{dv_0} = 1_N$. Note that for each time-step the derivatives $J_f(v_s)$ and $c(v_{s-1})$ have been previously computed in (4.24) during the application of Newton-Raphson's method to (4.23). In fact, $J_f(v_s)$ is available in LU factored form. However, dv_{s-1}/dv_0 is a full $N \times N$ matrix that must be multiplied by both $c(v_{s-1})/h_s$ and $J_f^{-1}(v_s)$ (this last multiplication is done with N sparse forward and backward substitutions) at every time-step. Thus, (4.26) represents a burdensome calculation that prevents Newton-Raphson-based shooting methods from being applied to large circuits.

The size of $J_\phi(v_0, 0, T)$ can be reduced somewhat by eliminating entries in v from consideration that result from algebraic constraints or from quickly decaying states. This approach has also been found to aid convergence [kakizaki85].

Since (4.20) is being solved using Newton-Raphson, it is necessary to assure that $\phi(v_0, 0, T)$ is continuously differentiable with respect to v_0. Sufficient conditions for this to be true before time has been discretized are that i, q and u satisfy the conditions imposed in Section 1.3 to assure that a unique solution to (3.23) exists for every initial state, as well as the additional conditions that i and q are continuously differentiable with respect to v [hale80]. These conditions are also sufficient to assure that $\phi(v_0, 0, T)$ is continuously differentiable with respect to v_0 when time has been discretized with a fixed discretization mesh. However, $\phi(v_0, 0, T)$ is not guaranteed to vary smoothly with changes in the mesh. Indeed, insertion or deletion of points is inherently a discontinuous operation that is likely to introduce convergence problems if a course mesh is used unless care is taken. It is possible to avoid this problem by simply using the same mesh on each iteration, however this may result in unacceptable constraints on the time-step-selection algorithm and therefore either excessive truncation error[1] or an excessively fine mesh. To avoid these problems, a mesh is chosen to achieve the desired accuracy and used on successive iterations for as long as that accuracy can be maintained. Once it is no longer maintainable, the Newton iteration is restarted with a new mesh and using the final iterate from the previous mesh as the starting point. It may happen that when far from the solution, the mesh will be changed often, perhaps on each iteration. Once near the solution, however, the mesh should stabilize and allow the Newton iteration to converge.

Like any Newton-Raphson-based method, the shooting method will converge if the function (4.20) varies smoothly in the neighborhood of the solution and if the initial guess is given sufficiently close to the solution. Thus, an important part of the shooting method is the selection of a good initial guess for the solution. A reasonable start is to set the time-varying input sources to their average value and use the

---

[1]Truncation error is the error generated in discretizing a differential equation [gear71]. The truncation error is normally reduced by reducing the size of the time-steps.

resulting DC operating point for $v_0^{(0)}$. An improvement on this procedure is to linearize the circuit about the DC operating point, convert
the input source waveforms into the frequency domain using the
discrete Fourier transform, and perform a phasor analysis [desoer69] to
find the periodic steady-state response of the linearized circuit. From
this response, determine $v_0^{(0)}$ (this only requires the inverse discrete
Fourier transform to be evaluated at $t = 0$), and use this initial guess to
start the shooting method. The initial guess can be further improved
by integrating the circuit for several periods before applying the shooting method to allow any rapid time constants to decay. When generating the initial guess the truncation error criteria can be relaxed and
larger time-steps taken to reduce the computational cost.

It is still possible for the shooting method to have convergence
problems. As always with Newton-Raphson, damping can sometimes
be used to improve convergence. A damping factor α is introduced
into (4.21) so that

$$v_0^{(j+1)} = v_0^{(j)} + [\alpha J_\phi(v_0^{(j)}, 0, T) - D]^{-1}[v_0^{(j)} - \phi(v_0^{(j)}, 0, T)]$$

(4.27)

where $0 \le \alpha \le 1$. With $\alpha = 1$, (4.27) reverts to undamped Newton-
Raphson; if $\alpha = 0$ then (4.27) becomes the fixed-point iteration
$v_0^{(j+1)} = \phi(v_0^{(j)}, 0, T)$. The damping parameter should be automatically
controlled on each iteration with α close to zero when $\phi(\cdot, 0, T)$ is
strongly nonlinear over the size of a step $v_0^{(j+1)} - v_0^{(j)}$, and with α
close to one when $\phi(\cdot, 0, T)$ is almost linear over the step. To achieve
quadratic convergence, α must go to one as the solution is
approached. See [grosz82, kakizaki85] for specific algorithms to control α. Note that, for this damped shooting method to work, it is
necessary for $v_0^{(j)}$ to be within the region of attraction for an
asymptotically-stable limit cycle of period T, otherwise the fixed-point
iteration that results when $\alpha = 0$ will not converge.

A very important characteristic of shooting methods is that they
converge quickly and reliably if the state-transition function over the
shooting interval is near linear. It is quite often the case (usually by
design) that $\phi(\cdot, 0, T)$ is linear even when the overall circuit behavior is

not. For example, consider the simple one-stage switched-capacitor filter shown in Figure 4.1. Assume $v_0 = 0$ and C_{in} is connected to the input source at $t = 0^+$, the switches are thrown connecting C_{in} in parallel with C_f. After everything settles, $v_0 = (V_{in} C_{in})/(C_{in} + C_f)$. Before the end of the clock cycle the switches are returned to their original positions. If the shooting interval is taken to be one clock cycle, then $\phi(v_0, 0, T) = v_0 + (V_{in} C_{in})/(C_{in} + C_f)$. The state-transition function over the shooting interval is linear, even though the circuit might behave quite nonlinearly during certain portions of the clock cycle (ex. slew rate limiting). Shooting methods hide this nonlinear behavior from the outer loop, and so few iterations at this level are required. The nonlinear behavior is not a problem for the numerical integration used to evaluate $\phi(\cdot, 0, T)$ because numerical integration is a natural continuation method where time is the continuation parameter.

2.3. Oscillators

It is possible to handle oscillators with extrapolation-based shooting methods [skelboe80], however there is no natural way to find T. It is

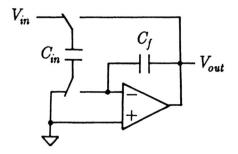

Figure 4.1: A simple one-pole switched-capacitor filter.

necessary to look for the maxima or threshold crossings on the solution waveforms and estimate T. Other than the methods that need be employed to find T, extension of extrapolation to the oscillator problem is straight-forward and well-behaved.

When applying Newton-Raphson-based shooting methods to autonomous oscillators, T is added to the list of unknowns to be determined and an extra equation is added that addresses the problem of having a continuum of solutions. The added equation is constructed either to eliminate almost all solutions (with those that remain being isolated from each other) or to modify Newton-Raphson so that it is capable of handling the continuum. One way to restrict the set of solutions such that each is isolated is to force $v(0)$ to lie on a hyperplane that has been carefully selected so that it intersects the solution trajectory. The resulting system of equations for a periodic oscillation is

$$\phi(\hat{v}(0), 0, T) - \hat{v}(0) = 0 \qquad (4.28)$$

$$\xi^T \hat{v}(0) = \alpha$$

where ξ is a constant vector that is normal to the hyperplane and α is a scalar. Unfortunately, ξ and α cannot be chosen arbitrarily, they must satisfy

$$\max_t \xi^T \hat{v}(t) > \alpha$$

$$\min_t \xi^T \hat{v}(t) < \alpha$$

$$\xi^T \frac{d\hat{v}(0)}{dt} \neq 0.$$

As shown in Figure 4.2, the first two constraints assure that the hyperplane intersects the solution orbit. The last constraint is necessary to prevent the Jacobian from being singular, which results if the hyperplane is tangent to the solution trajectory at $\hat{v}(0)$. Initially, ξ and α are selected after integrating the circuit equations for a few cycles, as are the initial Newton guess of T and $v(0)$. It may be necessary to update α as the iteration progresses.

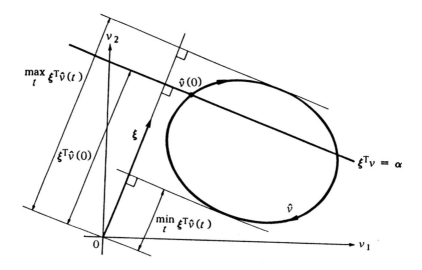

Figure 4.2: Orbit and hyperplane of Eqn. (4.28). Shows in two dimensions how $v(0)$ is selected by intersecting a hyperplane with the solution trajectory.

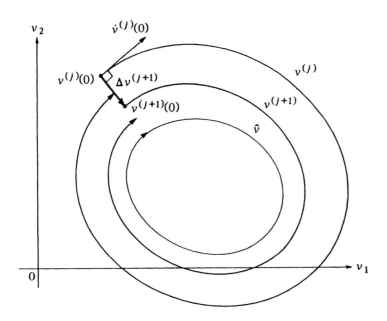

Figure 4.3: Two steps of the shooting method on a two dimensional oscillator problem with $\Delta v^{(j+1)}$ constrained to be perpendicular to $\dot{v}^{(j)}(0)$.

Mees modified Newton-Raphson-based shooting [mees81] to handle nonisolated solutions by constraining the Newton-step direction to be perpendicular to the trajectory, as shown in Figure 4.3. Thus if $\Delta v^{(j+1)}(0) = v^{(j+1)}(0) - v^{(j)}(0)$, then the equation $\dot{v}(0)^{\mathrm{T}}\Delta v(0) = 0$ is added to the Newton update iteration for $\phi(v(0), 0, T) - v(0) = \mathbf{0}$ where both $v(0)$ and T are considered unknown.

$$\begin{bmatrix} J_\phi(v^{(j)}(0), 0, T^{(j)}) - \mathbf{1} & \dfrac{\partial\phi(v^{(j)}(0), 0, T^{(j)})}{\partial T} \\ \dot{v}^{(j)}(0)^{\mathrm{T}} & 0 \end{bmatrix} \begin{bmatrix} \Delta v^{(j+1)}(0) \\ \Delta T^{(j+1)} \end{bmatrix}$$

$$= \begin{bmatrix} v^{(j)}(0) - \phi(v^{(j)}(0), 0, T^{(j)}) \\ \mathbf{0} \end{bmatrix} \qquad (4.29)$$

where

$$J_\phi(v(0), 0, T) = \frac{\partial \phi(v(0), 0, T)}{\partial v(0)} \in \mathbb{R}^{N \times N}$$

$$\frac{\partial \phi(v(0), 0, T)}{\partial T} \in \mathbb{R}^N$$

$$\Delta T^{(j+1)} = T^{(j+1)} - T^{(j)}.$$

This approach does not require the selection of parameters such as ξ and α, but a standard Newton-Raphson package cannot be used because the added equation is in terms of $\Delta v(0)$ rather that $v(0)$. Small errors in $\dot{v}^{(j)}(0)$ do not affect the accuracy of the solution, and so it can be calculated using the approximation $\dot{v}(0) \approx (v(t_1) - v(t_0))/h_1$. The computation of $\partial \phi(v(0), 0, T)/\partial T$ is more difficult. As before, the derivative must be derived from the discretized circuit equations. Equation (3.11) is discretized using implicit Euler and $\partial \phi(v(0), 0, T)/\partial T = \partial v_S/\partial T$ is computed by taking the final value of the $\partial v_s/\partial T$ trajectory. To compute $\partial v_s/\partial T$, it is necessary to specify how t_s, $s = 0, 1, \ldots, S$ vary with T. Let $t_s = \alpha_s T$ where $0 = \alpha_0 < \alpha_1 < \cdots < \alpha_S = 1$. From (4.23)

$$\frac{1}{h_s}\left[q(v_s) - q(v_{s-1})\right] + i(v_s) + u_s = \mathbf{0}$$

where $s = 1, 2, \ldots, S$, $v_s = v(\alpha_s T)$ and $u_s = u(\alpha_s T)$. Differentiating with respect to T leads to

$$\frac{1}{h_s}\left[\frac{\partial q(v_s)}{\partial T} - \frac{\partial q(v_{s-1})}{\partial T} - \frac{q(v_s) - q(v_{s-1})}{T}\right] + \frac{\partial i(v_s)}{\partial T} + \frac{\partial u_s}{\partial T} = \mathbf{0}.$$

Applying the chain rule and recalling that u is constant gives

$$\frac{1}{h_s}\left[\frac{\partial q(v_s)}{\partial v_s}\frac{\partial v_s}{\partial T} - \frac{\partial q(v_{s-1})}{\partial v_{s-1}}\frac{\partial v_{s-1}}{\partial T} - \frac{q(v_s) - q(v_{s-1})}{T}\right] +$$

$$\frac{\partial i(v_s)}{\partial v_s} \frac{\partial v_s}{\partial T} = 0.$$

Again let $g = \partial i / \partial v$ and $c = \partial q / \partial v$.

$$\left[g(v_s) + \frac{c(v_s)}{h_s} \right] \frac{\partial v_s}{\partial T} = c(v_{s-1}) \frac{\partial v_{s-1}}{\partial T} + \frac{1}{h_s T} \left[q(v_s) - q(v_{s-1}) \right]$$

$$\frac{\partial v_s}{\partial T} = J_f^{-1}(x_s) \left[c(v_{s-1}) \frac{\partial v_{s-1}}{\partial T} + \frac{1}{h_s T} \left[q(v_s) - q(v_{s-1}) \right] \right] \qquad (4.30)$$

This difference equation is solved starting with the initial state $\frac{\partial v_0}{\partial T} = 0$.

It is important to realize that Newton-Raphson has been applied to the set of finite-difference equations that approximate the original set of differential equations, and not directly to the differential equations. In computing the derivatives of v_S needed by Newton-Raphson, it is therefore necessary to use the finite-difference equations. It is also important to notice that the time-points are independent variables that affect the value of v_S. If the time-points are not fixed, they must be included in the list of independent variables being solved for by the Newton iteration. This has been done in a manner that is at least tractable by allowing T to be independent and insisting that $t_s = \alpha_s T$, $s = 0, 1, \ldots, S$. However, this approach constrains severely the time-steps and may result in excessive truncation error. If the truncation error becomes intolerable, a new mesh is chosen with both S and the α_s's changing in such a way that the error specification is satisfied. In doing so, the finite difference equations are changed and Newton-Raphson should be restarted. However the last iterate from the previous mesh may be used as an initial guess for the iteration on the new mesh. Initially, the mesh may need to be changed often, but once near the solution the mesh should stabilize and allow the Newton iteration to converge.

It is possible to change the way in which $\partial v_S / \partial T$ is computed, which may result in some gain in efficiency. One way is to fix the time-steps $t_0, t_1, \ldots, t_{S-1}$, and only allow $t_S = T$ to vary. This greatly simplifies (4.30) but requires that the mesh be updated more often. Another approach is to use the approximation $\partial v_S / \partial T \approx \dot{v}(T)$. Thus, in an interesting twist, the original continuous time solution is used as an approximation to the discrete-time solution. When using this approach, it is necessary to use a fine enough mesh for the approximation to be accurate.

2.4. Distributed Devices

When applying finite-difference method to circuits with distributed devices, the approach taken was to use the impulse response to describe a distributed device and discretize the impulse response in time. In doing so, only the signals on the terminals of the distributed device were important, making it easy to characterize the device from measurements made at the terminals. The tradeoff was the signals on the terminals were needed for all past time. This tradeoff was acceptable when computing the steady-state solution with finite-difference methods because the solution is calculated for all time. However, with shooting methods the solution is not known for any time preceding the shooting interval. Thus, characterizing a distributed device with its impulse response is inappropriate and it is necessary instead to discretize the distributed device in space rather than in time. With shooting methods the initial state of the distributed device along its entire discretized length must be determined. One problem with distributed devices in shooting methods is that measurements made from the terminals are difficult to convert into a space-discretized model for the device.

Commonly, the distributed devices present in high frequency circuits can only be represented accurately by lumped models with a large number of nodes, primarily because these devices have very high delay-bandwidth products. For example, consider an ideal transmission line. A model for the ideal line is a series inductor / shunt capacitor ladder network. The values of inductance (L) and capacitance (C) per section would be chosen to provide the correct characteristic impedance and to make the bandwidth of the model (roughly equals $1/2\pi\sqrt{LC}$) larger than the frequencies expected in the circuit. The number of sections (n) is chosen to give the model the correct amount of delay $(td \approx n\sqrt{LC})$. For lines with considerable delay at the highest frequency of interest, a large number of sections are required. The computational complexity of both extrapolation- and Newton-Raphson-based shooting methods increases rapidly with the number of state variables. As a result, shooting methods are generally impractical for circuits containing more than one or two distributed devices.

2.5. Parallel Shooting

Shooting methods can have convergence and overflow problems when applied to circuits with unstable modes. For each iteration, only the initial state is specified, therefore any unstable modes may cause the trajectory to grow exponentially and leave the vicinity of the desired solution. The likelihood of this occurring decreases if the shooting interval T is reduced. Parallel, or multiple, shooting methods [keller68] [keller76] [stoer80] effectively reduce the shooting interval by dividing it into several subintervals. For example, if the shooting interval is divided into two equally sized subintervals, shooting is performed twice per iteration, once for each subinterval. Thus, the problem is restructured into another two-point boundary-value problem with twice as many unknowns (two sets of initial conditions, one set per subinterval) but with the shooting interval halved.

Consider dividing the interval $[0,T]$ for the lumped periodic test problem (3.11) into K subintervals defined by the mesh $\{\tau_0, \tau_1, \tau_2, \ldots, \tau_K\}$ where $\tau_0 = 0$, $\tau_K = T$ and $\tau_k < \tau_{k+1}$. Let $v_k = v(\tau_k)$ and define $\phi_k(v_k) = v_{k+1}$. Since v is T-periodic, $v_0 = v_K$, and by forcing v to be continuous at each τ_k

$$\phi_0(v_0) - v_1 = \mathbf{0}$$

$$\phi_1(v_1) - v_2 = \mathbf{0}$$

$$\vdots \qquad \vdots \qquad \vdots$$

$$\phi_{K-1}(v_{K-1}) - v_0 = \mathbf{0} \, .$$

$$(4.31)$$

This is a system of nonlinear algebraic equations that is solved using Newton-Raphson.

$$\begin{bmatrix} J_{\phi_0}(v_0^{(j)}) & & & -I \\ & J_{\phi_1}(v_1^{(j)}) & \ddots & \\ & & \ddots & -I \\ -I & & & J_{\phi_{K-1}}(v_{K-1}^{(j)}) \end{bmatrix} \begin{bmatrix} v_0^{(j+1)} - v_0^{(j)} \\ v_1^{(j+1)} - v_1^{(j)} \\ \vdots \\ v_{K-1}^{(j+1)} - v_{K-1}^{(j)} \end{bmatrix}$$

$$= \begin{bmatrix} v_1^{(j)} - \phi_0(v_0^{(j)}) \\ v_2^{(j)} - \phi_1(v_1^{(j)}) \\ \vdots \\ v_0^{(j)} - \phi_{K-1}(v_{K-1}^{(j)}) \end{bmatrix} \qquad (4.32)$$

By taking this iteration to convergence the value of the solution waveform is found at each of the points of the mesh. Both $\phi_k(v_k)$ and $J_{\phi_k}(v_k)$ are evaluated by integrating the circuit equations as in the previous subsection.

The computational complexity of the parallel shooting method is higher than the standard shooting method because the Jacobian in (4.32) has much larger dimension than the one in (4.21). The Jacobian is treated as a sparse block matrix and is efficiently factored by simply avoiding operations on zero blocks and possibly by applying the Sherman, Morrison, Woodbury formula [householder75] to convert it into a banded matrix. Besides the disadvantage of increased computational complexity, parallel shooting methods are also more difficult to program and require considerably more memory. These disadvantages, however, are offset by two advantages. First, the parallel shooting methods are more suitable for implementation on parallel processing computers; a feature that should grow in importance in the future. A second minor advantage is that parallel shooting methods have better convergence properties on unstable circuits. In particular, the region of convergence is larger for the parallel shooting methods and increases in size as the number of subintervals increases [keller76].

In the parallel shooting method, the circuit equations must be integrated over each subinterval individually. Thus each subinterval is further subdivided to perform the numerical integration of the differential equations. As the number of shooting subintervals increases, the number of subdivisions per interval decreases. This causes a problem if it has been deemed desirable to use a high-order integration method. At each step a high-order integration method needs the history of the solution over several past time-steps. This history cannot extend beyond a shooting interval boundary. Thus it is necessary to build up to higher order integration methods by taking several steps of a low order method at the beginning of each interval. When an interval contains only a few time-points, high order methods loose their advantage because of the large percentage of time-steps taken with the low order methods.

Chapter 5
Harmonic Balance Theory

1. Introduction

Harmonic balance differs from traditional transient analysis in two fundamental ways. These differences allow harmonic balance to compute periodic and quasiperiodic solutions directly and in certain circumstances give the method significant advantages in terms of accuracy and efficiency. Transient analysis, which uses standard numeric integration, constructs a solution as a collection of time samples with an implied interpolating function. Typically the interpolating function is a low order polynomial. However, polynomials fit sinusoids poorly, and so many points are needed to approximate sinusoidal solutions accurately.

The first difference between harmonic balance and transient analysis is that harmonic balance uses a linear combination of sinusoids to build the solution. Thus, it approximates naturally the periodic and quasiperiodic signals found in a steady-state response. If the steady-state response consists of just a few dominant sinusoids, which is common, then harmonic balance needs only a small data set to represent the response accurately. The advantage of using sinusoids to approximate an quasiperiodic steady-state response becomes particularly important when the response contains dominant sinusoids at widely separated frequencies.

Harmonic balance also differs from traditional time-domain methods in that time domain simulators represent waveforms as a collection of samples whereas harmonic balance represents them using the coefficients of the sinusoids. (Just as in traditional time-domain methods where it is presumed that a polynomial is used to interpolate between samples, we can use samples to represent the combination of sinusoids, with the understanding that a sum-of-sinusoids interpolation is to be done between samples.) Representing signals with the coefficients is a practical matter, the solution calculated will be the same with either representation, but using coefficients makes it easier to model and evaluate the linear components. Working with the coefficients and exploiting superposition makes it possible to calculate symbolically the response from linear dynamic operations such as time integration, differentiation, convolution, and delay. Because linear devices respond at the same frequency as the stimulus, it is only necessary to determine the magnitude and phase of the response. Using phasor analysis [desoer69], this is easily done for lumped components such as resistors, capacitors and inductors; while it is not trivial for the more esoteric distributed devices like transmission lines with dispersion, it is generally much easier to find their response using phasor analysis than to try to determine their response to sampled waveforms in the time domain. The use of phasors to evaluate the linear devices is perhaps the most useful feature of harmonic balance.

The major difficulty with the harmonic balance approach is determining the response of the nonlinear devices. There is no known way to compute the coefficients of the response directly from the coefficients of the stimulus for an arbitrary nonlinearity, though it is possible if the nonlinearity is described by a polynomial or a power series [steer83] [rhyne88]. It is not necessary to consider only these special cases, nor to accept the error of using them to approximate arbitrary nonlinearities. Instead, we convert the coefficient representation of the stimulus into a sampled data representation; this is a conversion from the frequency domain to the time domain and is accomplished with the inverse Fourier transform. With this representation the nonlinear devices are easily evaluated. The results are converted back into coefficient form using the forward Fourier transform.

Because the coefficients of the steady-state response are an algebraic function of the coefficients of the stimulus, the dynamic aspect of the problem is eliminated. Thus, the nonlinear integro-differential equations that describe a circuit are converted by harmonic balance into a system of algebraic nonlinear equations whose solution is the steady-state response of the circuit. These equations are solved iteratively.

Harmonic balance was given its name because it was viewed as a method for balancing of currents between the linear and nonlinear sub-circuits. Furthermore, harmonic balance is usually considered a mixed-domain method, because the nonlinear devices are evaluated in the time domain while the linear devices are evaluated in the frequency domain. However, evaluating the nonlinear devices in the time domain is not a fundamental part of the algorithm, but rather a convenience that does not affect the essential character of the algorithm. It is the formulation of the circuit equations in the frequency domain that give harmonic balance its essential characteristics. Thus, harmonic balance can be summarized as just being the method where KCL is formulated in the frequency domain.

This chapter begins with a brief history of the development of the harmonic balance algorithm followed by a simple example which is used to illustrate the method and some of its error mechanisms. The harmonic balance equations are then derived for the distributed test problem. The resulting system of nonlinear equations is solved using nonlinear programming methods, nonlinear relaxation, and the Newton-Raphson algorithm. Finally, the magnitude of the errors present in harmonic balance are estimated and techniques to extend harmonic balance to autonomous circuits are discussed.

2. Historical Development

Harmonic balance has been around for many decades. It was originally considered an approximate technique for finding analytically the near-sinusoidal solution of a differential equation [cunningham58] (a solution where all components except for one sinusoid is negligible). Baily formulated harmonic balance as a numerical method in 1969 [baily69]. He described the nonlinearities using polynomials, formulated Kirchoff's laws in terms of the Fourier coefficients, and solved the resulting nonlinear equations with nonlinear programming techniques. In the last 15 years, it has been reformulated into an accurate method for finding numerically the solution of a differential equation driven by sinusoids without having to approximate the nonlinearities with polynomials. The conventional approach begins by partitioning the circuit into linear and nonlinear subcircuits. The linear subcircuit is evaluated in the frequency domain while the nonlinear subcircuit is evaluated in the time-domain. The problem then becomes one of finding the voltage waveforms (or spectra) on the nodes that appear in both subcircuits that result in Kirchoff's current law being satisfied at those nodes. In 1974, Egami showed that it is possible to solve these equations using Newton-Raphson [egami74]. He was quickly followed by Gwarek and Kerr, who solved the equations using nonlinear relaxation [gwarek74] [kerr75], and Nakhla and Vlach who used optimization methods [nakhla76]. Variations of these three approaches have been presented by a large collection of authors [gopal78] [filicori79] [faber80] [mees81] [hicks82b] [hicks82a] [lipparini82] [penalosa83] [rizzoli83] [ushida84] [rizzoli86] [hente86] [penalosa87b] [penalosa87a] [curtice87] [maas88].

The relaxation methods presented before 1986 were based on simple splitting of the linear and nonlinear portions of the Jacobian [faber80, hicks82b]. At that time, Kundert and Sangiovanni-Vincentelli explored the convergence properties of the various harmonic balance methods and showed that convergence properties of this method were poor. We proposed an alternative relaxation method called Gauss-Jacobi-Newton harmonic relaxation that has superior convergence properties [kundert86b, schuppert86].

Relaxation methods were found to be quite fast but suffered from convergence problems. Newton-Raphson, on the other hand, is more robust but is slow. In 1985, Kundert and Sangiovanni-Vincentelli [kundert85] [kundert86b] presented a method, referred to here as harmonic relaxation-Newton, that combines most of the advantages of both relaxation and Newton-Raphson. The method smoothly adapts from relaxation to Newton-Raphson at the component level as individual devices are driven harder and their nonlinear character becomes more apparent. In one limit, a purely linear circuit, the method becomes equivalent to Gauss-Jacobi-Newton harmonic relaxation, which converges in one inexpensive iteration. In the other limit, for portions of the circuit behaving very nonlinearly, the method becomes identical to Newton-Raphson. This approach provides the speed of the relaxation methods without the risk.

This work is unique in another way. One assumption common to all previous approaches is that the circuit should be partitioned into two subcircuits. With such an approach, the system of equations that describe the linear subcircuit can be factored into triangular form once and quickly evaluated thereafter. This is desirable when the linear subcircuit is much larger than the nonlinear subcircuit. This is invariably true for hybrid circuits. However, treating the circuit as two separated subcircuits places constraints on the way the nonlinear equations can be factored and so is inefficient for monolithic circuits, where the number of nonlinear devices is large. Kundert and Sangiovanni-Vincentelli were the first to suggest that the circuit should not be partitioned into two subcircuits. Instead, the system of equations for the whole circuit should be carefully factored in a manner that is efficient for both hybrid and monolithic circuits.

A significant departure from conventional harmonic balance was suggested by Steer and Kahn in 1983 [steer83] [rhyne88]. They propose to evaluate the nonlinear devices directly in the frequency domain. It is only possible to do this for selected nonlinearities, such as those described by polynomials (making the method reminiscent of that of Baily). Thus, when analyzing circuits containing nonlinearities described by arbitrary continuous functions, the first step is to model the functions over the anticipated operating range with a polynomial. This approach has the advantage of eliminating any Fourier transforms in the nonlinear device evaluations, however these transforms usually do not require much time, and so the gain in efficiency due to this feature is negligible. The disadvantage of this method is that it is very difficult to approximate accurately strongly nonlinear functions over a wide range with polynomials, thus the method should only be applied to circuits with mild nonlinearities.

2.1. Harmonic Balance for Quasiperiodic Signals

Until 1984, harmonic balance was only used to analyze circuits with a periodic response. The reason being is that with the linear devices evaluated in the frequency domain and the nonlinear devices evaluated in the time domain, a transform is needed to convert signals between the two domains. The Fast Fourier Transform (FFT) was used to perform this operation, however the FFT is only applicable to periodic signals, which excludes a very important class of circuits whose steady-state response is not periodic: mixers. The signals present in mixers consist of sinusoids at the sum and difference frequencies of the two or more input frequencies and their harmonics. In general, these input frequencies are not harmonically related, and so the signals found in mixers are not periodic, but rather quasiperiodic.

In 1984, Ushida and Chua showed that a transform for the quasiperiodic signals present in mixers could be developed by starting from the matrix form of the Discrete Fourier Transform (DFT) but that more than the normal number of time-points were needed in the sampled time-domain waveforms [ushida84]. The excess number of samples necessitated the use of least-squares to perform the transformation of signals to and from the time and frequency domains. This approach allowed harmonic balance to be applied to mixers, but was disappointing for two reasons. First, the extra time samples represented a computational burden, the need for which remained unexplained. Second, the normal equation was used to formulate the least-squares problem, which is notoriously ill-conditioned.

Also in 1984, Gilmore and Rosenbaum [gilmore84, gilmore86] presented a completely different transform that exploited sparsity in a spectrum. In theory, it was limited to periodic signals, but in practice, quasiperiodic signals can be approximated arbitrarily closely by periodic signals. This fact is not usually useful when using conventional transforms such as the DFT and the FFT because to approximate quasiperiodic signal accurately with a periodic signal it is usually necessary to use a very low fundamental frequency. Thus the ratio between the highest and lowest frequencies, and hence the number of frequencies needed, is very large. However, the signal at almost all harmonics is zero (i.e., the spectrum is sparse), and so Gilmore and Rosenbaum's transform still remains competitive. The transform samples the waveform using several small sets of equally spaced time-points. The DFT is applied to each set individually. The sets are too small to prevent aliasing in the computed spectra. The aliasing is eliminated by taking an appropriate linear combination of the computed spectra. The spectra are constrained to be periodic in this method because the DFT is used. Though this transform can be much more efficient than the standard DFT on sparse spectra, the total number of time-points used is normally greater than the theoretical minimum by about 50%.

In 1987, Sorkin, Kundert, and Sangiovanni-Vincentelli [kundert88d, sorkin87] showed that using equally spaced samples in the time domain leads to severe ill-conditioning in the transform that could only be remedied by either using more than the theoretical minimum number of time samples (as in the least squares approach above), or using unequally spaced samples. We presented an algorithm for finding a minimum set of unequally spaced time samples that yields a well-conditioned transform. Using that transform, deemed the Almost-Periodic Fourier Transform (APFT), we developed a very simple and theoretically useful derivation of the harmonic balance algorithm for both the periodic and quasiperiodic cases.

If harmonic balance is applied to a circuit containing algebraic nonlinearities, then it is possible to use the FFT rather than a transform designed for quasiperiodic signals. One such approach based on multidimensional FFTs was proposed by Bava et al [bava82] and later presented by Rizzoli [rizzoli87] and Ushida in 1987 [ushida87]. A slightly more flexible approach was developed independently by Hente and Jansen in 1986 [hente86] and by Kundert in 1987 that uses one-dimensional FFTs. There are two aspects of the second approach that make it very attractive, first is simply that the quasiperiodic algorithm can be made essentially identical to the periodic algorithm. The second reason is very subtle and involves the interaction of aliasing and convergence, but the end result is a considerable reduction in the computation time. While this transform appears to be the most attractive for use in harmonic balance, it seems to have been completely overlooked by all but its developers.

2.2. An Example

As an example of how harmonic balance can be used to find the solution to a nonlinear differential equation, consider Duffing's equation, which can be used to describe a nonlinear LC circuit.

$$\ddot{x} + \lambda^2 x + \mu x^3 = A_1 \cos(\omega_o t) \tag{5.1}$$

The "amount of nonlinearity" in the equation is controlled by μ, and λ is the resonant frequency of the circuit when $\mu = 0$. The periodic steady-state solution to this equation has the form $x = \sum\limits_{k=0}^{\infty} a_k \cos(k\omega_o t)$ where $a_k = 0$ for $k = 0, 2, 4, \cdots$. To make the problem tractable, only a_1 and a_3 will be assumed to be nonzero. Substitute the assumed solution $\hat{x}(t) = a_1\cos(\omega_o t) + a_3\cos(3\omega_o t)$ into (5.1).

$$[\tfrac{1}{4}\mu(3a_1^2 a_3 + 6a_1 a_3^2 + 3a_1^3) + (\lambda^2 - \omega_o^2)a_1]\cos(\omega_o t) +$$

$$[\tfrac{1}{4}\mu(3a_3^3 + 6a_1^2 a_3 + a_1^3) + (\lambda^2 - 9\omega_o^2)a_3]\cos(3\omega_o t) +$$

$$\tfrac{3}{4}\mu(a_1^2 a_3 + a_1 a_3^2)\cos(5\omega_o t) +$$

$$\tfrac{3}{4}\mu a_1 a_3^2 \cos(7\omega_o t) +$$

$$\tfrac{1}{4}\mu a_3^3 \cos(9\omega_o t) = A_1\cos(\omega_o t) \tag{5.2}$$

Using the orthogonality of sinusoids at different frequencies, rewrite (5.2) as a system of five equations, one for each harmonic generated by the assumed solution.

$\cos(\omega_o t)$:

$$\tfrac{1}{4}\mu(3a_1^2 a_3 + 6a_1 a_3^2 + 3a_1^3) + (\lambda^2 - \omega_o^2)a_1 = A_1 \tag{5.3a}$$

$\cos(3\omega_o t)$:

$$\tfrac{1}{4}\mu(3a_3^3 + 6a_1^2 a_3 + a_1^3) + (\lambda^2 - 9\omega_o^2)a_3 = 0 \qquad (5.3b)$$

$\cos(5\omega_o t)$:

$$\tfrac{3}{4}\mu(a_1^2 a_3 + a_1 a_3^2) = 0 \qquad (5.3c)$$

$\cos(7\omega_o t)$:

$$\tfrac{3}{4}\mu a_1 a_3^2 = 0 \qquad (5.3d)$$

$\cos(9\omega_o t)$:

$$\tfrac{1}{4}\mu a_3^3 = 0 \qquad (5.3e)$$

Since there are only two unknowns, it is not possible to satisfy all five equations exactly. This problem results from including only a finite number of harmonics in the assumed solution when really an infinite number exists. Traditionally, the coefficients of the sinusoids in the solution are computed by solving the equations at the harmonics present in the solution. Thus a_1 and a_3 are found by solving (5.3a) and (5.3b) simultaneously. In effect, the exact solution is found for (5.1) with a perturbed right-hand side.

$$\ddot{\hat{x}} + \lambda^2\hat{x} + \mu\hat{x}^3 = A_1\cos(\omega_o t) + A_5\cos(5\omega_o t) +$$

$$A_7\cos(7\omega_o t) + A_9\cos(9\omega_o t) \qquad (5.4)$$

where $A_5 = -\tfrac{3}{4}\mu(a_1^2 a_3 + a_1 a_3^2)$

$\quad A_7 = -\tfrac{3}{4}\mu a_1 a_3^2$

$\quad A_9 = -\tfrac{1}{4}\mu a_3^3$

Notice that no mention has been made about how to solve the system of algebraic equations generated in the last step of the method of harmonic balance. Several different approaches have been used, the most notable being optimization [filicori79, gopal78, nakhla76], non-linear relaxation [hicks82b], and Newton-Raphson [egami74] [ushida84]. All these methods have quite different characteristics, but all have been referred to only as harmonic balance, which has led to a certain amount of confusion. To eliminate any confusion, in this book, the three approaches are be referred to as harmonic programming, harmonic relaxation, and harmonic Newton.

3. Error Mechanisms

There are three sources of error that are of interest in harmonic balance. The first two result from truncating the harmonics considered to some finite number, and the third results from not completely converging the iteration used to solve the nonlinear system of algebraic equations. If Newton-Raphson is used, then the third source of error can be driven to an arbitrarily small level in relatively few iterations because of the method's quadratic convergence property. So this source of error will be ignored for now.

As shown in (5.4), harmonics that are not in the assumed solution end up perturbing the right-hand-side of the algebraic equations. Recall that (5.3a) and (5.3b) were solved exactly for a_1 and a_3 and (5.3c), (5.3d) and (5.3e) were left unsatisfied; thus (5.1) was also unsatisfied. Let ε be the amount by which (5.1) is not satisfied

$$\varepsilon(\hat{x}, t) = \ddot{\hat{x}} + \lambda^2 \hat{x} + \mu \hat{x}^3 - A_1 \cos(\omega_o t)$$

where

$$\hat{x}(t) = \hat{a_1} \cos(\omega_o t) + \hat{a_3} \cos(3\omega_o t)$$

From (5.4) it is clear that

$$\varepsilon(\hat{x}, t) = A_5 \cos(5\omega_o t) + A_7 \cos(7\omega_o t) + A_9 \cos(9\omega_o t)$$

Note that the residual ε is orthogonal to the form of the assumed solution $\alpha_1 \cos(\omega_o t) + \alpha_3 \cos(3\omega_o t)$, making harmonic balance a Galerkin method. This property is lost if the nonlinear devices are evaluated in the time domain and the transform used exhibits aliasing.

An iterative method is used to solve the nonlinear algebraic system of equations generated by harmonic balance. Equation (5.3) is an example of such a system; it can be represented as

$$F(\hat{X}) = 0$$

where $F:\mathbb{R}^2 \to \mathbb{R}^5$ and $\hat{X} = [\hat{a}_1 \ \hat{a}_3]^{\mathrm{T}}$. In practice, this system is evaluated at $X = [\alpha_1 \ \alpha_3]^{\mathrm{T}}$ by computing $x(t) = \alpha_1\cos(\omega_o t) + \alpha_3\cos(3\omega_o t)$ at several time-points $t_1, t_2, \ldots, t_S$; evaluating $f(t) = \ddot{x} + \lambda^2 x + \mu x^3 - A_1\cos(\omega_o t)$ at these time-points; and converting $f(t)$ into the frequency domain using the discrete Fourier transform (DFT). In this example, $f(t)$ is band-limited, so its Fourier coefficients can be calculated exactly. Only the coefficients of the first two harmonics of f are of interest, the remaining ones are discarded. However, since there are 9 harmonics present, the Nyquist sampling theorem states that f must be evaluated at more than 18 time-points to determine the coefficients for the first two harmonics accurately. Originally, when it was decided to compute only two harmonics, it was assumed that the coefficients at the remaining harmonics are small. So for efficiency, when calculating the Fourier series of f, the remaining harmonics are assumed to be negligible, and f is evaluated at only enough points to calculate the coefficients of the first two harmonics. Since the remaining harmonics are not zero, they will alias down onto the two, resulting in further error.

4. Derivation

Recall that the statement of Kirchoff's current law for the distributed test problem (3.12) is

$$f(v,t) = i(v(t)) + \dot{q}(v(t)) + \int_{-\infty}^{t} y(t - \tau)v(\tau)d\tau + u(t) = 0$$

When applying harmonic balance to this problem, both v and $f(v)$ are transformed into the frequency domain. Since v is quasiperiodic (by assumption), both $i(v)$ and $q(v)$ are quasiperiodic; therefore all three waveforms can be written in terms of their Fourier coefficients: $\mathbb{F}v = V$, $\mathbb{F}i(v) = \mathbb{F}i(\mathbb{F}^{-1}V) = I(V)$ and $\mathbb{F}q(v) = \mathbb{F}q(\mathbb{F}^{-1}V) = Q(V)$, where $\mathbb{F}$ represents the Fourier transform operator. Since v, $i(v)$ and $q(v)$ are vectors of waveforms — one waveform for each node in the circuit — V, $I(V)$ and $Q(V)$ are vectors of spectra. The Fourier coefficients of the convolution integral are computed by exploiting its linearity. Assume y satisfies

$$\int_{-\infty}^{\infty} y(t)^T y(t) dt < \infty,$$

and $y(t) = 0$ for all $t < 0$; that is, assume y is causal and has finite energy (or equivalently, that the circuit with all nonlinear devices removed is causal and asymptotically stable); then

$$\mathbb{F} \int_{-\infty}^{t} y(t - \tau)v(\tau)d\tau = YV$$

where

$$Y = [Y_{mn}] \qquad m, n = 1, 2, \ldots, N$$

$$Y_{mn} = [Y_{mn}(k, l)] \qquad k, l \in \mathbb{Z}$$

where m, n are the node indices; k, l are the frequency indices, and

$$Y_{mn}(k, l) = \begin{cases} \begin{bmatrix} \mathrm{Re}\{Y_{mn}(j\omega_k)\} & \mathrm{Im}\{Y_{mn}(j\omega_k)\} \\ -\mathrm{Im}\{Y_{mn}(j\omega_k)\} & \mathrm{Re}\{Y_{mn}(j\omega_k)\} \end{bmatrix} & \text{if } k = l \\ 0 & \text{if } k \neq l \end{cases}$$

where Y is the Laplace transform of y [desoer69] and $j = \sqrt{-1}$.

Now (3.12) can be rewritten in the frequency domain as

$$F(V) = I(V) + \Omega Q(V) + YV + U = 0 \qquad (5.5)$$

where $U = \mathbb{F}u$ contains the Fourier coefficients for the source currents over all nodes and frequencies, and

$$\Omega = [\Omega_{mn}] \qquad m, n = 1, 2, \ldots, N$$

$$\Omega_{mn} = \begin{cases} [\Omega_{mn}(k, l)] & \text{if } m = n \\ 0 & \text{if } m \neq n \end{cases}$$

$$\Omega_{mn}(k,l) = \begin{cases} \begin{bmatrix} 0 & \omega_k \\ -\omega_k & 0 \end{bmatrix} & \text{if } k = l \\ 0 & \text{if } k \neq l \end{cases}.$$

That $\mathbb{F}\dot{q}(v) = \Omega Q(V)$ follows from the differentiation rule of the Fourier series. Equation (5.5) is simply the restatement of Kirchoff's current law in the frequency domain.

It is important to realize that the frequency-domain functions for the nonlinear devices (I and Q) are evaluated by transforming the node voltage spectrum V into the time domain, calculating the response waveforms i and q, and then transforming these waveforms back into the frequency domain. To assure that the nonlinear device response waveforms are quasiperiodic, we require that the nonlinear devices be algebraic. If not (that is, if the device has memory), then the response waveform has a transient component, is not quasiperiodic, and cannot be accurately transformed into the frequency domain. The restriction that nonlinear devices be algebraic clearly allows nonlinear resistors. Fortunately, it also allows nonlinear capacitors and inductors (actually, any lumped nonlinear component) because their constitutive relations are algebraic when written in terms of the proper variables; v and i for resistors, v and q for capacitors, and i and ϕ for inductors [chua80]. The conversion between i and q ($i = dq/dt$) and v and ϕ ($v = d\phi/dt$) is done in the frequency domain where it is an algebraic operation and does not disturb the steady-state nature of the solution. Nonlinear distributed devices, however, are not algebraic, and the trick of evaluating their response in the time domain and transforming it into the frequency domain cannot be used. Instead, it is necessary to remain in the frequency domain and model the nonlinear device using a Volterra series representation or develop an approximate lumped model for the device. Conventional transient analysis is also not able to handle nonlinear distributed devices, a mixed device / circuit simulator such as CODECS [mayaram88] or MEDUSA [engl82] is needed. Fortunately, the transmission lines used in high-frequency analog and microwave design are all linear, even those that are dispersive or lossy. We will not consider nonlinear distributed devices further.

5. Harmonic Programming

It is possible to apply nonlinear programming techniques to solve (5.5). To do so use $\varepsilon(V) = \frac{1}{2}F^{\mathrm{T}}(V)F(V)$ as the cost function where $\varepsilon(V) \in \mathbb{R}$. An optimizer is used to find the value of V that globally minimizes $\varepsilon(V)$. If a $\hat{V}$ is found such that $\varepsilon(\hat{V}) = 0$, then $\hat{V}$ satisfies (5.5).

Applying nonlinear programming to solve the harmonic balance equations is expensive because there is a very large number of unknowns to be found and because nonlinear programming techniques are expensive for large problems. If a circuit with 20 nodes is simulated at 8 harmonics, then 300 variables need to be optimized. If there are many nodes in the circuit that have only linear devices attached, then it is possible to shrink the number of variables to be optimized by considering the collection of all linear devices as a single multiterminal subcircuit. The nodes with no nonlinear devices attached become internal nodes to the subcircuit and so need not be considered as optimization variables. Figure 5.1 shows a convenient way of visualizing the analysis once the linear devices have been placed in a subcircuit. Here, the substitution theorem has been used to replace the nonlinear devices with sources. The resulting circuit is linear, however, the voltage spectra for the voltage sources that are used to replace the nonlinear devices are unknown. These spectra are generated by the optimization package. Nakhla and Vlach [nakhla76] have taken this idea one step further by considering the collection of nonlinear devices as a subcircuit as well. Neither of these two approaches help when MMIC's are simulated however because each node in a monolithic circuit tends to have linear devices as well as nonlinear resistors and capacitors tied to them.

Using an optimizer to solve the harmonic balance equations is inefficient. To do so requires that a difficult problem, that of solving $F(V) = 0$ for V, is converted into an even harder problem, that of solving $\dfrac{dF^{\mathrm{T}}(V)F(V)}{dV} = 0$. That information is lost in the conversion aggravates the situation. All information about each of the individual contributors to ε is lost when $F^{\mathrm{T}}(V)F(V)$ is formed to calculate $\varepsilon(V)$.

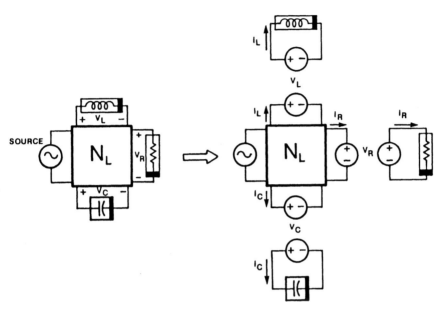

Figure 5.1: Circuit interpretation of harmonic programming.

It is also difficult to exploit the structure of the harmonic balance / node admittance equations. For these reasons, the approaches presented later are preferred over harmonic programming.

The next section presents how to solve the harmonic balance equations with optimization techniques and the reasons why this is not the preferred approach. The situation where it is desired to do performance optimization while solving the harmonic balance equations is also discussed.

5.1. Root Finding with Nonlinear Least Squares

To use optimization methods to solve (5.5), it is necessary to develop an appropriate cost function that has minima at the same values of V for which F has roots. The most commonly used cost function results from formulating the problem as a nonlinear least-squares problem,

$$\varepsilon(V) = \tfrac{1}{2} F^T(V) F(V). \tag{5.6}$$

This cost function has two important characteristics. First, each root of F corresponds to a global minimum of ε, and at these points, $\varepsilon(V) = 0$. Second, at each V that is a local minima of ε but is not a root of F, the Jacobian[1] of F (that is $J_F(V) = dF(V)/dV$) is singular. We seek $\hat{V}$, a global minimizer of ε. A necessary condition for $\hat{V}$ to be a minimizer of ε is that the gradient[2] of ε at $\hat{V}$ be zero, i.e.,

$$\nabla\varepsilon(\hat{V}) = 0. \tag{5.7}$$

This problem can be solved by using a wealth of techniques such as steepest descent, conjugate gradient, and Newton's method. The steepest descent algorithm has the advantage of having a region of convergence that is usually much larger than the other methods, however it has only a linear rate of convergence and often converges so slowly as to be impractical. The conjugate gradient algorithm has a superlinear rate of convergence, but for large problems it usually requires a large number of iterations. Each iteration is inexpensive, and so this might be a suitable algorithm for solving the harmonic balance equations. Newton's method uses Newton-Raphson[3] to find the roots of (5.7). Thus with Newton's method we are again faced with using Newton-

---

[1] A Jacobian is the first derivative of a vector-valued function with respect to its vector-valued argument.

[2] A gradient is the first derivative of a scalar-valued function with respect to its vector-valued argument.

[3] There can be considerable confusion when discussing optimization methods and root-finding methods simultaneously because both the root-finding method and the optimization method that results from applying Newton-Raphson to find the roots of the gradient are referred to as Newton's method. In this book, the root-finding procedure is referred to as Newton-Raphson and the optimization method as Newton's method.

Raphson method to find the root of a nonlinear equation. However, solving (5.7) with Newton-Raphson is more difficult than solving (5.5) because the equation involves the first derivative of the original function F, so applying Newton-Raphson requires knowing the second derivatives of F. In fact, the Newton-Raphson iteration used to solve (5.7) is

$$\nabla^2 \varepsilon(V^{(k)})[V^{(k+1)} - V^{(k)}] = -\nabla\varepsilon(V^{(k)}), \qquad (5.8)$$

where $\nabla^2 \varepsilon$ is the Hessian[4] of ε.

From (5.6), it is easy to show that

$$\nabla\varepsilon(V) = J_F^T(V)F(V) \qquad (5.9)$$

and

$$\nabla^2\varepsilon(V) = \frac{dJ_F(F)}{dV}F(V) + J_F^T(V)J_F(V). \qquad (5.10)$$

Clearly, the Hessian is denser than is the Jacobian and is therefore considerably more expensive to LU factor. Using (5.9) and (5.10), we can rewrite (5.8) as

$$\left[\frac{dJ_F(V)}{dV}F(V) + J_F^T(V)J_F(V)\right][V^{(k+1)} - V^{(k)}] = -J_F^T(V)F(V)$$

$$(5.11)$$

Traditionally, computing the Hessian is avoided by using quasi-Newton methods [gill81], which build up an approximation to the Hessian by starting with an identity matrix and performing rank-one updates computed from changes in the gradient at each step. Quasi-Newton methods require more iterations because they use an approximate Hessian, and so more gradient evaluations, but avoid computing the Hessian directly. This approach is interesting because it is possible to form the Cholesky[5] factors of the approximate Hessian and perform the

---

[4]A Hessian is the second derivative of a scalar-valued function with respect to its vector-valued argument.

[5]It is possible to write a symmetric positive-definite matrix (such as the Hessian)

rank-one updates on these factors directly. Thus each iteration is considerably less expensive than evaluating (5.11) directly, but a greater number of iterations are needed. If compared against solving (5.5) directly with Newton-Raphson, quasi-Newton methods are at a disadvantage because the Hessian is denser than the Jacobian and because the Hessian can become ill-conditioned when $F(V)$ is small, as will be explained below.

It is possible to avoid computing the second derivative terms in the Hessian by exploiting the fact that we are solving for a root of F. When V is near the solution $\hat{V}$, then $F(V)$ should be small, and so the second derivative term $[dJ_F(V)/dV \; F(V)]$ in (5.11) can be dropped. Thus (5.11) becomes

$$J_F^{\mathrm{T}}(V)J_F(V)[V^{(k+1)} - V^{(k)}] = -J_F^{\mathrm{T}}(V)F(V), \qquad (5.12)$$

which is referred to as the Gauss-Newton algorithm. Note that (5.12) is now in the form of the normal equation for a linear least-squares problem. This equation is ill-conditioned because the condition number of $J_F^{\mathrm{T}}(V)J_F(V)$ is the square of $J_F(V)$. Thus, finding the roots of $F(V)$ using (5.12) is not only computationally more expensive (because $J_F^{\mathrm{T}}(V)J_F(V)$ is denser than $J_F(V)$), but it is also possibly numerically unstable.

By assuming that $J_F(V)$ is square and nonsingular, it is possible to cancel $J_F^{\mathrm{T}}(V)$ from both sides of (5.12), resulting in the following iteration,

$$J_F(V)[V^{(k+1)} - V^{(k)}] = -F(V). \qquad (5.13)$$

This iteration is identical to that which results when Newton-Raphson is applied to (5.5). However, by assuming that $J_F(V)$ is nonsingular, any possibility of finding a minimum of ε that is not a root of F has been eliminated. (Recall that minima occur where either F is zero or where J_F is singular.) Thus, (5.13) is not an optimization method, but rather a root finding method. Indeed, (5.13) is identical to the

---

as a product or a lower triangular matrix and its transpose. These triangular matrices are the Cholesky factors of the original matrix [strang80]

Newton-Raphson iteration, and so it can have no hidden powers not shared by Newton-Raphson. In fact, the steps that lead up to (5.13) are just a rather round-about derivation of the harmonic Newton algorithm that is presented later.

The ill-conditioning demonstrated in Gauss-Newton is an inherent consequence of forming the cost function by squaring F. Any optimization algorithm that begins with (5.6) will suffer from greater numeric instability than results from simply finding the root of F using Newton-Raphson. In addition, most of the sparsity and the structure of the Jacobian is missing from the Hessian. Methods that use the Hessian must labor under this disadvantage. Optimization methods that do not use the Hessian, such as the conjugate gradient algorithm, require a much greater number of iterations, precisely because they do not use the Hessian. Since these iterations are relatively inexpensive, these methods may be useful for solving the harmonic balance equations, but this has yet to be shown.

5.2. Performance Optimization

An interesting question remains; if it is necessary to both optimize circuit performance and solve the harmonic balance equations, should these two operations be combined into one optimization process? This idea was first suggested by Lipparini, et al [lipparini82]. They proposed to augment the cost function to be minimized with a contribution related to circuit performance, here denoted E. The list of design parameters is denoted p. The problem statement becomes

$$\min_{V,p} [F^{\mathrm{T}}(V,p)F(V,p) + E^2(V,p)] \qquad (5.14)$$

with the added constraint that $E = 0$ when all specifications are met and $E > 0$ otherwise. This approach appears attractive because simple unconstrained optimization methods are used.

Unfortunately, a serious problem exists: (5.14) is not an appropriate statement of the problem. The correct problem statement is

$$\min_{p} E^2(V,p) \qquad \text{subject to} \qquad F(V,p) = 0 \qquad (5.15)$$

By using (5.14), we allow the optimizer trade off satisfying Kirchoff's current law to improve circuit performance. The flaw, of course, is that if Kirchoff's current law is not satisfied, the solution calculated is not feasible and therefore the actual circuit performance is not being measured.

It is possible to solve (5.15) directly using a variety of methods, including the method of Lagrange multipliers, exact penalty function methods, and Lagrangian methods [bertsekas82] [luenberger84]. Lipparini's approach should not be used unless something is done to assure that Kirchoff's current law is satisfied. One possibility is to weight Kirchoff's current law so that it is given preference over the performance goals. This weight can be increased as necessary to assure that Kirchoff's current law is sufficiently satisfied. Such an approach is known as a penalty function approach, which is known to be inefficient when the penalty function (here the weighted Kirchoff's current law) is large.

Assuming something can be done to assure that Kirchoff's current law is satisfied, then there are several other important considerations. To treat (5.14) as a nonlinear least squares problem it is necessary to augment the list of equations with the performance cost function to be minimized and augment the list of variables with the optimizable parameters. Let

$$X = \begin{bmatrix} V \\ p \end{bmatrix} \qquad\qquad G(X) = \begin{bmatrix} F(X) \\ E(X) \end{bmatrix}. \qquad (5.16)$$

The cost function to be minimized becomes

$$G^{\mathrm{T}}(X)G(X). \qquad (5.17)$$

Augmenting the lists of equations and variables presents several problems. First, the new equations in G create new rows and columns in the Jacobian J_G that do not have the same structure as in J_F, making exploitation of the sparsity of the Jacobian more difficult. Second, there is usually more than one design parameter in p. Thus, J_G is not square. When applying Newton-Raphson to a system with more variables than equations, it is necessary to solve for the new iterate either by forming the normal equation, which effectively increases the number of equations until equal to the number of unknowns, or to solve the iteration equation with a method that is suitable for under-determined systems, such as QR factorization. The normal equation approach is written as

$$J_G^T(X^{(k)})J_G(X^{(k)})[X^{(k+1)} - X^{(k)}] = -J_G^T(X^{(k)})G(X^{(k)})$$

(5.18)

As mentioned before, this equation is ill-conditioned and not nearly as sparse as when Newton-Raphson is applied directly to (5.5). QR factorization would be applied to

$$J_G(X^{(k)})[X^{(k+1)} - X^{(k)}] = -G(X^{(k)}).$$ (5.19)

However, it is not possible to exploit sparsity in any significant way using QR factorization, and so this approach is impractical for large $J_G(X)$.

6. Harmonic Relaxation

Relaxation methods are another approach that can be used to solve the algebraic set of equations that result from the application of harmonic balance. These methods are attractive when the nonlinear behavior of the circuit is very mild. Two different ways of applying relaxation methods are presented, the first uses a form of nonlinear relaxation called splitting that is similar to the approach taken by Gwarek or Kerr [gwarek74] [kerr75] [hicks82b] [faber80]. The second combines relaxation and Newton-Raphson; it has much better convergence properties than the first approach [kundert86b] [ushida87].

6.1. Splitting

Splitting is a relaxation technique that was originally developed to solve linear systems of equations and was generalized to handle non-linear systems [ortega70]. As an introduction, consider the linear system

$$Ax = b$$ (5.20)

and consider the splitting of A into the sum

$$A = B - C$$

where B is nonsingular and the system $Bx = d$ is easy to solve. Then a fixed-point iteration that can be applied to find the solution of (5.20) is

$$x^{(j+1)} = B^{-1}(Cx^{(j)} + b)$$

where the superscript on x is the iteration count. This iteration will converge if all the eigenvalues of $B^{-1}C$ are smaller in magnitude than one.

The splitting method can be used with harmonic balance by rewriting equation (5.5) as

$$YV^{(j+1)} = -I(V^{(j)}) - \Omega Q(V^{(j)}) - U \qquad (5.21)$$

Y is assumed to be nonsingular, which implies that when all nonlinear devices are removed, there can be no floating nodes. Since linear devices are incapable of translating frequencies, the node admittance matrix for the linear portion of the circuit (Y) is block diagonal (we are assuming here that the harmonic number is the major index and the node number is the minor index). Thus once the right-hand side of (5.21) has been evaluated, the task of finding $V^{(j+1)}$ can be broken into solving K decoupled linear $N \times N$ systems of equations, one for each harmonic.

To explore the convergence properties of the iteration defined by (5.21), the following well-known theorem [ortega70] [rudin76] [vidyasagar78] from classical analysis is needed.

Theorem 5.1 *(Contraction Mapping Theorem): Let C be a closed subset of $\mathbf{C}^M$. If f is a map from C into C and if there exists $\gamma < 1$ such that*

$$\| f(x) - f(y) \| \le \gamma \| x - y \|$$

for all $x, y \in C$, then f is called a contraction map on C. Furthermore there exists a unique $\hat{x}$ (called a fixed point of f) such that $f(\hat{x}) = \hat{x}$ and given any $x^{(0)} \in C$, the sequence $\{x^{(j)}\}$ defined by

$x^{(j+1)} = f(x^{(j)})$ *converges to* $\hat{x}$.

☐

 If $C = \mathbf{C}^{KN}$, then the theorem gives sufficient conditions for the global convergence of (5.21), however it is difficult to apply, so a theorem giving sufficient conditions for local convergence is presented. But first a lemma is needed.

Lemma: *Suppose f maps a convex open set E contained in* $\mathbf{C}^M$ *into* $\mathbf{C}^M$, *f is differentiable in E, and there is a real number c such that* $J_f(x)$, *the Jacobian of f at x, satisfies*

$$\| J_f(x) \| \le c \quad \text{for every } x \in E$$

Then $\| f(x) - f(y) \| \le c \| x - y \|$ *for all* $x, y \in E$.

☐

 The lemma is a straight-forward extension of a theorem given by Rudin [rudin76] for $\mathbf{R}^N$.

Theorem 5.2: *Let E be an open subset of* $\mathbf{C}^M$. *Suppose* $f : E \to \mathbf{C}^M$ *is continuously differentiable on E and can be written in the form*

$$f(x) = Ax - g(x)$$

where $A \in \mathbf{C}^{M \times M}$ *is nonsingular. If there exists* $\hat{x} \in E$ *such that* $f(\hat{x}) = 0$ *and if* $\| A^{-1} J_g(\hat{x}) \| < 1$ *then there exists some* $\delta > 0$ *such that for all* $x^{(0)}$ *in the closed ball* $B_\delta(\hat{x}) \subset E$ *the sequence* $\{x^{(j)}\}$ *defined by* $x^{j+1} = A^{-1} g(x^{(j)})$ *converges to* $\hat{x}$.

Proof: Define $\phi(x) = A^{-1} g(x)$ and assume there exists some $\gamma \in [0,1)$ such that $\| A^{-1} J_g(\hat{x}) \| < \gamma$. Since f, and hence g, is continuously differentiable, there exists $\delta > 0$ such that for all $x \in B_\delta(\hat{x})$, $\| A^{-1} J_g(x) \| \le \gamma$. Note that the derivative of $\phi(x)$ is $J_\phi(x) = A^{-1} J_g(x)$. From the lemma,

$$\| \phi(x) - \phi(y) \| \le \gamma \| x - y \| \tag{5.22}$$

for all x, y in the interior of $B_\delta(\hat{x})$, and since ϕ is continuous, (5.22) must hold for all of $B_\delta(\hat{x})$, not just the interior. By the contraction mapping theorem, ϕ has a unique fixed-point in $B_\delta(\hat{x})$, which must be $\hat{x}$ because it is a fixed-point for ϕ. Also, $\{x^{(j)}\} \to \hat{x}$ if $x^{(0)} \in B_\delta(\hat{x})$

□

   If the conclusion of Theorem 5.2 is applied to (5.21), then to assure local convergence we need

$$\| Y^{-1}[J_I(\hat{V}) + \Omega J_Q(\hat{V})] \| < 1 \qquad (5.23)$$

where $\hat{V}$ is the solution of (5.21). There is no reason to believe this condition will be met in practice. As an example of when it would not be met, consider a very simple circuit with only one node and analyzed at DC only. Then $Y \in \mathbb{R}$ and $I, Q : \mathbb{R} \rightarrow \mathbb{R}$. Let $Y = 1$, $I(V) = 2V$, and $Q(V) = 0$. Then $Y^{-1}J_I = 2$ and so the convergence criterion is not satisfied. Indeed, for this circuit convergence will not be achieved for any $V^{(0)} \neq \hat{V}$. This example shows that relaxation using the splitting method given by (5.21) has poor convergence properties. In particular, even if the starting value of $V^{(0)}$ is arbitrarily close to the final solution, and regardless of how mild the nonlinearities are behaving, convergence is not assured. In fact, convergence can easily be lost if the largest conductance exhibited by any of the nonlinear devices is larger than the smallest conductance to ground present in the circuit when the nonlinear devices are removed.

6.2. Gauss-Jacobi-Newton Harmonic Relaxation

The second relaxation approach to solving the algebraic harmonic balance equation (5.5) is to use the block Gauss-Jacobi method with a one step Newton-Raphson inner loop [ortega70] [newton83] known as the block Gauss-Jacobi-Newton method. To apply this method, rewrite (5.5) as

$$F(V,k) = I(V,k) + \Omega(k,k)Q(V,k) + Y(k,k)V(k) + U(k) = 0 \quad (5.24)$$

where k is the frequency index, $k = 0, 1, \ldots, K-1$, and

$$\Omega(k,k) = [\Omega_{mn}(k,k)] \qquad m, n = 1, 2, \ldots, N$$

$$\Omega_{mn}(k,k) = \begin{bmatrix} 0 & \omega_k \\ -\omega_k & 0 \end{bmatrix}.$$

The block Gauss-Jacobi algorithm, when applied to (5.24), has the following form:

Nonlinear Block Gauss-Jacobi Algorithm

repeat
{ $j \leftarrow j + 1$;
     forall $(k \in \{0, 1, \ldots, K-1\})$
     {   **solve**
              $F(V^{(j)}(0), \ldots, V^{(j+1)}(k), \ldots, V^{(j)}(K-1), k)$
          for
              $V^{(j+1)}(k)$

     }
} **until** $(\| V^{(j+1)} - V^{(j)} \| < \varepsilon)$

The equation inside the **forall** loop is solved using Newton-Raphson for $V^{(j+1)}(k)$. Note that in this equation only $V(k)$ is a variable, $V(l)$ $l \neq k$ are constant and taken from the previous iteration. Since Newton-Raphson (with quadratic convergence) is being performed inside an outer relaxation loop (with linear convergence), it is not necessary to fully converge the Newton iteration. In fact, it is only necessary to take one step of the "inner" Newton iteration and doing so does not affect the asymptotic rate of convergence of the overall method [ortega70] [newton83].

Block Gauss-Jacobi-Newton is similar to the splitting method, except that each equation in (5.24) is first linearized with one step of Newton-Raphson rather than by just removing the nonlinearities. Applying the Gauss-Jacobi-Newton method to (5.24) results in

$$\frac{\partial F(V^{(j)},k)}{\partial V(k)} \left[V^{(j+1)}(k) - V^{(j)}(k) \right] = -F(V^{(j)},k)$$

where $k = 0, 1, \ldots, K-1$ and

$$\frac{\partial F(V,k)}{\partial V(k)} = \frac{\partial I(V,k)}{\partial V(k)} + \Omega(k,k)\frac{\partial Q(V,k)}{\partial V(k)} + Y(k,k)$$

To continue the derivation, it is necessary to select a set of time-points and then use the APFT (3.7) to develop a concrete representation Γ of the Fourier transform $\mathbb{F}$. Only the derivation of $\dfrac{\partial I(V,k)}{\partial V(k)}$ is presented, the derivation of $\dfrac{\partial Q\,(V,k)}{\partial V\,(k)}$ is identical.

$$I_m(V) = \Gamma i_m(v)$$

where m is the node index. Let $\rho(k) \in \mathbf{C} \times \mathbb{R}^S$ be the two rows of Γ that compute the sine and cosine terms of the kth harmonic of the Fourier series. That is

$$I_m(V,k) = \rho(k)^{\text{T}} i_m(v).$$

Differentiating both sides of $I_m(k)$ with respect to $V_n(k)$,

$$\frac{dI_m(V,k)}{dV_n(k)} = \rho(k)^{\text{T}} \frac{di_m(v)}{dV_n(k)}.$$

Employing the chain rule gives

$$\frac{dI_m(V,k)}{dV_n(k)} = \rho(k)^{\text{T}} \frac{di_m(v)}{dv_n} \frac{dv_n}{dV_n(k)}.$$

Since $i(v)$ is algebraic,

$$\frac{di_m(v)}{dv_n} = \left[\frac{di_m(v(t_r))}{dv_n(t_s)} \right]_{r,s \,\in\, \{1,2,\,\ldots\,,S\}} \in \mathbb{R}^{S \times S}$$

is a diagonal matrix. Let $\gamma(k) \in \mathbb{R}^S \times \mathbf{C}$ be the two columns of Γ^{-1} that operate on the sine and cosine terms of the kth harmonic of the Fourier series to compute their contribution to the waveform. Then $\Gamma^{-1}V_n = v_n$ implies $\sum\limits_{k=0}^{K-1} \gamma(k)V_n(k) = v_n$. Thus,

$$\frac{dI\,(V,k)}{dV_n(k)} = \rho(k)^{\text{T}} \frac{di_m(v)}{dv_n} \gamma(k).$$

The block Gauss-Jacobi-Newton iteration is well-defined only if each of the equations in (5.24) has a unique solution at each step. In addition, convergence can be assured at least in the region local to the solution $\hat{V}$, if certain conditions are met by the Jacobian $J_F(V)$ at the solution. In particular, if F is continuously differentiable on an open set $E \subset \mathbf{C}^{KN}$ containing $\hat{V}$, and if $J_F(V) = D(V) + R(V)$ where D is a nonsingular block diagonal matrix with $D(V,k,k) = \dfrac{\partial F(V,k)}{\partial V(k)}$, then from Theorem 5.2, there exists a closed ball $B_\delta(\hat{V}) \subset E$ such that for all $V^{(0)} \in B_\delta(\hat{V})$ the block Gauss-Jacobi iteration is well defined and will converge to $\hat{V}$ if $\| D^{-1}(\hat{V})R(\hat{V}) \| < 1$. Notice that $D(V,k,k)$ is the Jacobian of the circuit at the kth harmonic. In other words, it is the node admittance matrix of the circuit at the kth harmonic where the circuit has been linearized about the solution. R represents coupling between signals at different harmonics that results from nonlinearities in the circuit. If the circuit is linear then $R = 0$ and convergence is assured. The more strongly nonlinear the circuit behaves, the more coupling exists between different harmonics and the larger the terms in R become. Thus, convergence becomes less likely. So block Gauss-Jacobi-Newton is guaranteed to converge if $F(V)$ is "sufficiently linear" and if $V^{(0)}$ is sufficiently close to the solution $\hat{V}$.

This method can be generalized by allowing the block bandwidth of D to be increased. Doing so is the basis of the harmonic relaxation-Newton method presented in Chapter 6.

To illustrate how the two relaxation methods work, consider the network shown in Figure 5.2a. In the splitting method, on each iteration the voltages on the nonlinear devices are fixed at the values of the previous iteration, which fixes the current passed by these devices. So in Equation (5.21) the nonlinear currents are moved to the right-hand side with the constants, and in Figure 5.2 they are replaced with current sources, using the substitution theorem [desoer69]. This "linearizes" the circuit, so the node voltages can be found with Gaussian elimination. These new node voltages are then applied to the nonlinear devices, and their new current is calculated and applied to the linearized circuit, requiring it to be re-evaluated on the next iteration. The linearized circuit never changes, so only forward and backward substitution is needed for reevaluation.

With block Gauss-Jacobi-Newton the circuit is linearized by dividing the nonlinear devices into two parts. One is the best linear approximation to the nonlinear device considering the signal present on the device. The other is the nonlinear residual that when combined

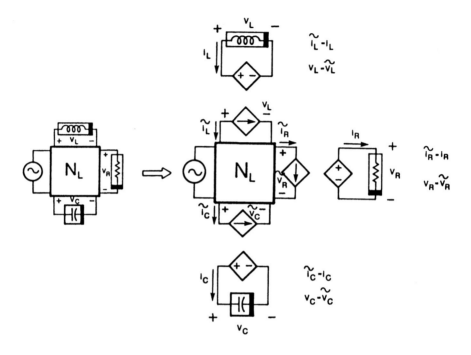

Figure 5.2: Circuit interpretation of the splitting method form of harmonic relaxation.

with the linear part gives the original nonlinear device. This division is illustrated in Figure 5.3.

By viewing Figures 5.2 and 5.3, it becomes clear why Gauss-Jacobi-Newton has better convergence properties than the splitting method; it has a better model of the nonlinear device in the linear sub-circuit, so less correction is needed on each iteration. Indeed, if the nonlinear devices behave linearly, then clearly Gauss-Jacobi-Newton

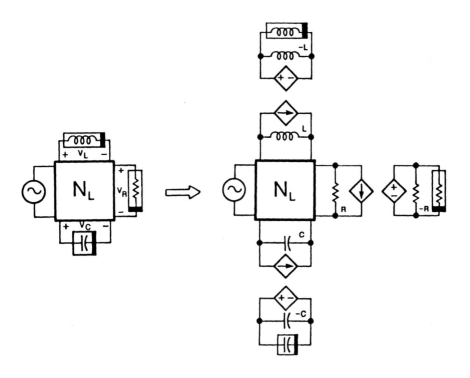

Figure 5.3: Circuit interpretation of the block Gauss-Jacobi-Newton form of harmonic relaxation.

converges in one step while the splitting will require many, and may not even converge.

The Gauss-Jacobi-Newton method has the nice feature that it uses very little memory. The circuit is analyzed at only one frequency at a time, so space is needed for only one sparse $N \times N$ node admittance matrix and that space is reused for each frequency. (This contrasts with the harmonic Newton method presented next. It analyzes the circuit at all frequencies simultaneously, and so it needs a great deal more memory.) The Gauss-Jacobi-Newton method is also quite fast if the circuit is behaving linearly. However it does have severe convergence problems when circuits behave nonlinearly. Harmonic Newton, which can be seen as a logical extension of Gauss-Jacobi-Newton, is much more robust, but also can require much more time on near linear circuits. For this reason, the harmonic relaxation-Newton method, as presented later, modifies harmonic Newton on near linear problems to become much more like Gauss-Jacobi-Newton, resulting in a composite method that can be both fast and robust.

7. Harmonic Newton

As shown earlier, the circuit equation

$$f(v,t) = i(v(t)) + \dot{q}(v(t)) + \int_{-\infty}^{t} y(t-\tau)v(t)dt + u(t) = 0$$

(5.25)

can be written in the frequency domain as

$$F(V) = I(V) + \Omega Q(V) + YV + U = 0. \tag{5.26}$$

One approach to evaluating the nonlinear devices in (5.26) is to convert the node voltage spectrum V into the waveform v and evaluate the nonlinear devices in the time domain. The response is then converted back into the frequency domain. Assume that the number of frequencies has been truncated to K; $v, u \in AP^N(\Lambda_K; \mathbb{R})$; and that a set of time-points $\{t_0, t_1, \ldots, t_{2K-1}\}$ has been chosen so that Γ^{-1} is non-singular. Then $V_n = \Gamma v_n$, $I_n(V) = \Gamma i_n(v)$ and $Q_n(V) = \Gamma q_n(v)$.

Applying Newton-Raphson to solve (5.26) results in the iteration

$$J_F(V^{(j)})(V^{(j+1)} - V^{(j)}) = -F(V^{(j)}) \tag{5.27}$$

where

$$J_F(V) = \frac{\partial F(V)}{\partial V} = \frac{\partial I(V)}{\partial V} + \Omega \frac{\partial Q(V)}{\partial V} + Y.$$

Or

$$J_F(V) = \left[J_{F,mn}(V)\right] = \left[\frac{\partial F_m(V)}{\partial V_n}\right]$$

$$m, n \in \{1, 2, \ldots, N\}$$

where

$$\frac{\partial F_m(V)}{\partial V_n} = \frac{\partial I_m(V)}{\partial V_n} + \Omega_{mm}\frac{\partial Q_m(V)}{\partial V_n} + Y_{mn}.$$

$J_F(V)$ is referred to as the *harmonic Jacobian*. The matrix $\partial F_m/\partial V_n$, known as a *conversion matrix*, is the derivative of the function at node m with respect to the Fourier coefficients of the voltage at node n. The derivation of $\partial I_m/\partial V_n$ follows with help from the chain rule.

$$I_m(V) = \Gamma i_m(v)$$

$$\frac{\partial I_m(V)}{\partial V_n} = \Gamma\frac{\partial i_m(v)}{\partial v_n}\frac{\partial v_n}{\partial V_n}$$

Since $i(v)$ is algebraic, $\partial i_m/\partial v_n$ is a diagonal matrix. Using the fact that $\Gamma^{-1}V_n = v_n$,

$$\frac{\partial I_m(V)}{\partial V_n} = \Gamma\frac{\partial i_m(v)}{\partial v_n}\Gamma^{-1}$$

The derivation of $\partial Q_m/\partial V_n$ is identical. Now everything needed to evaluate (5.27) is available. If the sequence generated by (5.27) converges, its limit point is the desired solution to (5.26).

The most computationally expensive part of harmonic Newton is the factorization of J_F, a $(N \times 2K)$ by $(N \times 2K)$ sparse matrix. Šamanskii's method [ortega70] can be employed to reduce the computation time required by the harmonic Newton algorithm. In this approach the factored Jacobian from the previous iteration is simply reused for several iterations. This algorithm was presented in Chapter 3.

Harmonic Newton, as with any Newton-Raphson-based method, is only guaranteed to converge if the initial guess is close enough to

the solution. Thus, finding a good initial guess is a key issue in determining the likelyhood of convergence. For many circuits, a good initial guess is generated by linearizing the circuit about the DC operating
point, applying the stimulus, and performing a phasor (AC) analysis at
each frequency in Λ_K. If the initial guess generated in this manner
does not result in convergence, it is necessary to use a variant of
Newton-Raphson that is more robust. Currently, the algorithm that
appears to be best suited in this situation is arc-length continuation as
presented in Appendix C.

One last feature of the harmonic Newton algorithm is that, once a
solution has been found, the resulting harmonic Jacobian represents a
linearization. Thus, if harmonic Newton is used to find a steady-state
solution, the Jacobian is the linearization of the circuit about its time-
varying operating point. Just as the Jacobian that results from a DC
analysis is the linearization of the circuit about its quiescent operating
point that is used in small signal calculations such as the AC and noise
analyses, the harmonic Jacobian can be used in small signal applications as well [held78, kerr79]. Since it is represents a time-varying
linear circuit, it exhibits frequency conversion.

8. Error Estimation

There are three dominant sources of error with harmonic balance. The
first results from incompletely converging the iteration used to solve
the nonlinear harmonic balance equations. This form of error was discussed in Chapter 3. The second form of error in harmonic balance
results from limiting the number of frequencies in the Fourier series
that represents the solution. This error has been explored by Huang
[huang89]. It is difficult to estimate, and so will not be discussed
further here.

The last form of error with harmonic balance results from using a
finite number of frequencies in the Fourier series representing $F(V)$.
The error can be split conceptually into two parts, truncation error and
aliasing. Consider a circuit whose exact solution is almost periodic
over the set of frequencies Λ and consider finding an approximate solution by applying harmonic balance on the truncated set of frequencies

$\Lambda_K \subset \Lambda$. Thus, given any $V \in \mathbf{C}^{NK}$, let $F : \mathbf{C}^{NK} \to \mathbf{C}^{NK}$ be the sum of the currents at every node and at K frequencies as in (5.5). For the purposes of the error analysis, several variants of F are defined to be the sum of the currents at every node and every frequency. Let $F_{full} : \mathbf{C}^{N\infty} \to \mathbf{C}^{N\infty}$ be the result when the circuit equations are evaluated without error at all frequencies in Λ. Let $F_{trunc} : \mathbf{C}^{N\infty} \to \mathbf{C}^{N\infty}$ be defined as

$$F_{trunc}(V,k) = \begin{cases} 0 & \text{if } \omega_k \in \Lambda_K \\ F_{full}(V,k) & \text{otherwise.} \end{cases}$$

Thus, F_{trunc} represents all currents generated by the circuit at frequencies other than those in Λ_K. These currents are not explicitly handled by discrete Fourier transforms (any discrete Fourier transform, including the DFT and APFT) and so these currents are mistaken for currents at frequencies in Λ_K. These error currents, referred to as aliasing, are represented by $F_{alias} : \mathbf{C}^{N\infty} \to \mathbf{C}^{N\infty}$. Finally, let $F_{approx} : \mathbf{C}^{N\infty} \to \mathbf{C}^{N\infty}$ be defined as

$$F_{approx}(V,k) = \begin{cases} F(V,k) & \text{if } \omega_k \in \Lambda_K \\ 0 & \text{otherwise.} \end{cases}$$

These four quantities are related by

$$F_{full} = F_{approx} - F_{alias} + F_{trunc}$$

To estimate the effect of truncation and aliasing errors, let $\hat{V} \in \mathbf{C}^{N\infty}$ be the exact solution and $\tilde{V} \in \mathbf{C}^{N\infty}$ be the solution in the presence of truncation and aliasing errors. Expand

$$F_{full}(\hat{V}) = 0$$

into a Taylor series about the solution to compute ΔV, the estimated difference between these two solutions.

$$F_{full}(\hat{V}) + \frac{\partial F_{full}(\hat{V})}{\partial V} \Delta V \approx F_{full}(\tilde{V})$$

$$F_{full}(\hat{V}) + \frac{\partial F_{full}(\hat{V})}{\partial V}\Delta V \approx F_{approx}(\tilde{V}) + F_{trunc}(\tilde{V}) - F_{alias}(\tilde{V})$$

$$0 + \frac{\partial F_{full}(\hat{V})}{\partial V}\Delta V \approx 0 + F_{trunc}(\tilde{V}) - F_{alias}(\tilde{V})$$

$$\frac{\partial F_{full}(\hat{V})}{\partial V}\Delta V \approx F_{trunc}(\tilde{V}) - F_{alias}(\tilde{V})$$

This equation gives an estimate of the error in the solution, but in terms of quantities that are unknown. $F_{trunc}(\tilde{V})$ and $F_{alias}(\tilde{V})$ will be estimated below. With harmonic Newton, $\partial F_{full}(\hat{V})/\partial V$ could be approximated by $J_F(\tilde{V})$, but since $J_F(\tilde{V}) \in \mathbf{C}^{NK \times NK}$, the effect of F_{trunc} would be lost. This is not a serious problem because $\| F_{alias} \|$ is generally as large or larger than $\| F_{trunc} \|$. Better approximations to $\partial F_{full}(\hat{V})/\partial V$ can be easily constructed by either treating the circuit as linear at frequencies not in Λ_K, or by exploiting the Toeplitz / Hankel structure of the conversion matrices (presented in Chapter 6) to extend them to frequencies outside of Λ_K.

One simple approach to estimating the error due to truncation and aliasing is to simply increase the number of frequencies used in the Fourier series representation of $F(V)$. For example, using the same solution, reevaluate all of the nonlinear devices at, say, $2S$ rather than the normal S samples, and compute the Fourier series of the resulting waveforms. One note of caution, when using this approach with the APFT, one must be very careful when choosing the new larger set of sample times to assure that error due to aliasing is actually reduced.

It is possible to effectively double the number of frequencies in the Fourier series of the response of a nonlinear device without reevaluating the device at further time-points. Instead, use the previously computed response waveform at the original S samples and make some further calculations to compute the time derivative of the waveforms at these samples. By using the slope of the waveform at each sample along with the value, twice as much information is available, and so the Fourier series can be computed with twice as many terms. For example, consider a nonlinear resistor with current

$i = i(v(t))$ and conductance $di(v(t))/dv(t) = g(v(t))$. The voltage waveform v is given at S distinct time-points, and the response waveform is computed by simply evaluating the resistor current equation at these points. The time derivative is computed using the chain rule

$$\frac{di(v(t))}{dt} = \frac{di(v(t))}{dv(t)} \frac{dv(t)}{dt}$$

$$\dot{i}(v(t)) = g(v(t))\dot{v}(t)$$

The conductance g is available because it was needed for the Newton-Raphson iteration and $\dot{v}(t)$ is easily computed by transforming ΩV into the time domain. With the current (or charge) and its time derivative known at each of the S sample points, the Fourier series is computed using the ordinate and slope discrete Fourier transform [bracewell78].

The method of estimating the error given above is an inexpensive way of determining whether enough frequencies have been included in the harmonic balance analysis after a solution has been computed. Huang has developed an apriori estimate of the number of frequencies needed to achieve a prespecified accuracy [huang89].

9. Oscillators

Recall that oscillators present two problems not found with forced circuits. The period of the oscillation is unknown and must be determined, and the time origin is arbitrary and thus if one solution exists, then an infinite continuum of solutions exists. In other words, if v is a solution, then so is any time shifted version of v. The problem is that Newton-Raphson fails if the solution is not isolated. It is necessary to modify either Newton-Raphson to handle nonisolated solutions or the problem formulation to eliminate the nonisolated solutions.

Harmonic balance is modified to handle oscillators by adding the fundamental frequency ω to the list of unknowns and an equation to enforce the constraint that solutions be isolated from one another. Perhaps the easiest way to ensure isolated solutions is to chose some signal in the circuit and insist that the sine part of its fundamental is zero. This fixes the solution to within a sign change. For this approach to work it is necessary that the fundamental of the signal chosen have nonzero magnitude.

The harmonic balance equations modified for the oscillator problem are

$$F(V,\omega) = I(V) + \Omega(\omega)Q(V) + Y(\omega)V + U = 0 \qquad (5.28)$$

$$V_m^S(1) = 0$$

Newton's method is now applied to these two equations,

$$\begin{bmatrix} J_F(V^{(j)},\omega^{(j)}) & \dfrac{\partial F(v^{(j)},\omega^{(j)})}{\partial \omega} \\ e_m^S(1)^T & 0 \end{bmatrix} \begin{bmatrix} \Delta V^{(j+1)} \\ \Delta \omega^{(j+1)} \end{bmatrix} = - \begin{bmatrix} F(V^{(j)},\omega^{(j)}) \\ V_m^S(1) \end{bmatrix}$$

$$(5.29)$$

where

$$\frac{\partial F(v,\omega)}{\partial \omega} = \frac{\Omega(\omega)}{\omega}Q(V) + \frac{\partial Y(\omega)}{\partial \omega}V$$

and $e_m^S(1)$ is the unit vector that selects the sine part of the first harmonic of the chosen node voltage.

The most difficult part of applying harmonic balance to the oscillator problem is determining a good initial guess for both V and ω. A good initial guess is required not only for the standard reason of assuring convergence, but also to avoid the valid but undesirable solution where all but the DC terms in V are zero.

Chapter 6
Implementing Harmonic Balance

This chapter describes practical algorithms for implementing harmonic balance. Most of these algorithms have been implemented and tested in *Spectre*, a simulation program suitable for analog and microwave circuits. There are two fundamental ideas presented in this chapter. The first is that it is possible to use the DFT with harmonic balance, even when the signals present are not periodic. The benefit of the DFT is the very regular manner in which aliasing occurs, which can be exploited to accelerate the construction and factorization of the conversion matrices. The first portion of this chapter is dedicated to explaining how the DFT can be used in quasiperiodic harmonic balance, and rederiving harmonic Newton with the DFT.

The second idea is that ignoring terms in the Jacobian converts harmonic Newton into a relaxation process. Indeed, the new method that results if any contribution to the harmonic Jacobian due to nonlinear behavior is ignored is identical to harmonic relaxation based on splitting of (5.21). If instead, the off diagonal entries in each conversion matrix were set to zero, the resulting method would be Gauss-Jacobi-Newton harmonic relaxation. A new method, referred to as harmonic relaxation-Newton, is developed by ignoring terms in the Jacobian that are sufficiently small. This algorithm can be viewed as either the harmonic Newton algorithm with an approximate Jacobian or as a hybrid of the Gauss-Jacobi-Newton harmonic relaxation and harmonic Newton algorithms.

1. Accelerating Harmonic Balance

Of the time spent performing harmonic Newton, most is spent constructing and factoring the harmonic Jacobian $J_F(V)$. Forming $\Gamma \dfrac{\partial i_m(v)}{\partial v_n} \Gamma^{-1}$ requires $O(K^3)$ operations because of the matrix multiplies. There is one such product to form for each of the nonlinear conductors and capacitors. As the number of nonlinear components is typically $O(N)$, forming the Jacobian requires $O(NK^3)$ operations. The computational complexity of LU factoring the block Jacobian matrix is $O(N^\alpha K^3)$, where typically $1.1 < \alpha < 1.5$.[1] Clearly reducing the computation in both forming and factoring the Jacobian are important to improving the efficiency of harmonic Newton.

The algebraic nature of the nonlinearities can be exploited to reduce the time required to form the Jacobian. Algebraic nonlinearities allow the use of the DFT (or FFT) in lieu of the APFT,. This gives three benefits, the actual transform itself is faster (assuming the FFT is used), the construction of the conversion matrices is accelerated because it becomes possible to use an algorithm that requires $O(K^2)$ operations rather than $O(K^3)$, and the factorization of the conversion matrices can be accelerated by techniques given later that increase its sparsity. It is worth noting that in general, only a small fraction of the time required by the harmonic balance algorithm is spent executing the Fourier transform, even when the DFT is used. The benefit of using the FFT is minor (over the DFT or APFT) in comparison to the benefit of using the FFT or DFT (over the APFT) when constructing and factoring the conversion matrices. As an interesting aside, it may be more efficient to use the DFT than the FFT. The reason being that with the FFT the number of frequencies is constrained, usually to be a power of two and the DFT allows an arbitrary number of frequencies. In general, the number of frequencies needed with the FFT is at least as great

---

[1]This is an approximation. The true operation count for a sparse LU factorization, ignoring the cost of factoring the blocks, is $c_3 N^3 + c_2 N^2 + c_1 N$, where c_1, c_2, and c_3 are functions of the sparsity pattern of the matrix. Generally for the sizes and densities found in circuit matrices, the c_2 and c_3 terms are small enough so that the N^2

as with the DFT. While the extra frequencies can be discarded, the FFT will require the nonlinear devices to be evaluated at a greater number of time-points.

The time required to factor the conversion matrices can be considerably reduced by reducing the density of the conversion matrices by using a hybrid of the harmonic Newton and harmonic relaxation methods. Harmonic Newton can be converted to Gauss-Jacobi-Newton harmonic relaxation by simply setting all off-diagonal terms in the conversion matrices to zero. In the hybrid method, which is referred to as harmonic relaxation-Newton, the bandwidth of selected conversion matrices is reduced. In one limit (when operating on a linear circuit), the bandwidth of all conversion matrices is reduced until they are diagonal, the so the method becomes equivalent to the Gauss-Jacobi-Newton harmonic relaxation. In the other limit (when each device in the circuit is behaving very nonlinearly), the bandwidth of all conversion matrices is expanded until the whole matrix is included, and the method becomes equivalent to harmonic Newton. The bandwidth of the conversion matrices is chosen to be as small as possible without sacrificing convergence. The bandwidth of each conversion matrix on each iteration can be set independently to achieve this goal. In this way, the method adapts to the problem being solved. The method can be modeled as either a relaxation method or the Newton-Raphson method with an approximate Jacobian.

2. Quasiperiodic Harmonic Balance Transforms

There are currently five different methods available for transforming signals between time and frequency domains that are suitable for use with quasiperiodic harmonic balance. The first three of these methods are general in nature. Two methods, those of Ushida and Chua [ushida84] and Gilmore and Rosenbaum [gilmore84], were outlined in Chapter 5. They are considered less efficient than the remaining methods and are no longer used. The third method, the APFT, was presented in Chapter 3 and Appendix B. This transform has a simple

---

and N^3 terms are noticeable but do not dominate.

operator notation and so is useful for theoretical manipulation.

The last two methods exploit the fact that in harmonic balance one desires the frequency domain response of a nonlinear device to a frequency domain stimulus and the time-domain waveforms generated along the way are of no interest. These methods transform spectra into waveforms with a distorted time axis. However, they also convert the waveforms with a distorted time axis back into the spectra, and so as long as the nonlinearities being evaluated in the time domain are algebraic, which was a basic assumption with harmonic balance, then the resulting spectra are correct. Of these two remaining methods, the first is based on the multidimensional DFT [bava82, ushida87] and is restricted to the box truncation. The second is based on the one dimensional DFT. It is faster and has fewer restrictions than the multidimensional DFT approach, and is the one presented here.

2.1. Harmonic Balance and the DFT

There is a trick that allows the use of the DFT — and hence the FFT — with harmonic balance even when the desired solution is quasiperiodic. To use the trick, two conditions must be satisfied.

1. The nonlinearities must be algebraic and time invariant.

2. The signals must be quasiperiodic. This is not a limitation because circuits of practical interest are excited using a finite number of periodic input signals and so their response is quasiperiodic.

3. The time-domain signals must be of no interest. This is true in harmonic balance where going into the time domain is an expedient but not essential way to compute the frequency-domain response of the nonlinear devices to a frequency-domain stimulus.

The trick is best explained with an example. Consider a nonlinear resistor with the constitutive equation

$$i(v) = v^2.$$

Assume that this resistor is being driven with the voltage waveform

$$v(t) = \cos(\alpha t) + \cos(\beta t).$$

The resistor responds with a current waveform of

$$i(v(t)) = 1 + \tfrac{1}{2}\cos(2\alpha t) + \cos(\alpha t - \beta t) + \cos(\alpha t + \beta t) + \tfrac{1}{2}\cos(2\beta t).$$

Notice that the coefficients of the cosines (i.e., the spectrum of the response signal) are independent of the frequencies α and β. This is true, whenever the nonlinearities are algebraic. Thus, for the purposes of evaluating the nonlinear devices, the actual fundamental frequencies are of no importance and can be chosen freely. In particular, the fundamentals can be chosen to be multiples of some arbitrary frequency so that the resulting signals will be periodic. Once the fundamentals are chosen in this manner, the DFT can be used. It is important to realize that these artificially chosen fundamental frequencies are not actually used in the harmonic balance calculations, such as in Y or Ω. Rather, they are used when determining in which order to place the terms in the spectra so that the DFT can be used when evaluating the nonlinear devices.

2.1.1. Choosing the Artificial Frequencies

For simplicity, the way in which the artificial frequencies are chosen is illustrated by examples. The actual artificial frequencies, and the scale factors that convert the original fundamental to the artificial frequencies, are of no interest except in determining the correspondence between the quasiperiodic and periodic harmonic indices.

Consider the following set of frequencies.

$$\Lambda_K = \{\omega : \omega = k_1\lambda_1 + k_2\lambda_2; \; 0 \le k_1 \le H_1, \; |k_2| \le H_2,$$

$$k_1 \ne 0 \text{ if } k_2 < 0\}$$

Let $\alpha_1 = 1$ and $\alpha_2 = \lambda_1/[\lambda_2(2H_2 + 1)]$ be the scaling factors of the two fundamentals. Then the resulting scaled set of frequencies is equally spaced and no two frequencies overlap. This scaling, which is ideal for box truncations, is illustrated in Figure 6.1. The correspondence between original frequencies and artificial frequencies is given by

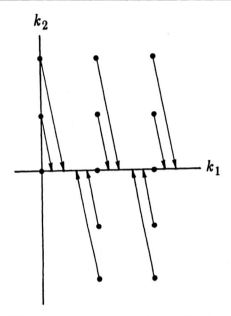

Figure 6.1: The mapping of quasiperiodic frequencies into periodic frequencies for box truncations.

$$k\lambda_0 = k_1\alpha_1\lambda_1 + k_2\alpha_2\lambda_2 \qquad (6.1)$$

where

$$\lambda_0 = \frac{\lambda_1}{2H_2 + 1}$$

and

$$k = (2H_2 + 1)k_1 + k_2$$

Consider another set of frequencies.

$$\Lambda_K = \{\omega : \omega = k_1\lambda_1 + k_2\lambda_2; \; |k_1| + |k_2| \le H, \; k_1 + k_2 \ge 0,$$

$$k_1 \ne k_2 \text{ if } k_2 > 0\}$$

Let $\alpha_1 = 1$ and $\alpha_2 = (\lambda_1 + H_2)/[\lambda_2(2H_2 + 1)]$ be the scaling factors of the two fundamentals. Then the resulting scaled set of frequencies is equally spaced and no two frequencies overlap. This scaling, which is ideal for diamond truncations, is illustrated in Figure 6.2. The correspondence between original frequencies and artificial frequencies is given by

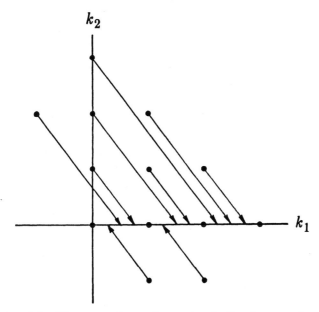

Figure 6.2: The mapping of quasiperiodic frequencies into periodic frequencies for diamond truncations.

$$k\lambda_0 = k_1\alpha_1\lambda_1 + k_2\alpha_2\lambda_2 \tag{6.2}$$

where

$$\lambda_0 = \frac{\lambda_1}{H_2 + 1}$$

and

$$k = (H_2 + 1)k_1 + H_2 k_2$$

Both approaches can be extended to the case where more than two fundamentals are applied. And while these two methods are ideal for their respective truncations, other truncations can be used, however the resulting set of frequencies may not be densely packed.

3. Harmonic Newton

The harmonic Newton algorithm is now rederived using the DFT rather than the APFT. It is assumed that, for the purposes of evaluating the nonlinear devices, the fundamentals have been shifted to make the signals processed by the DFT periodic.

There are two common DFT pairs that can be used in this derivation. Up to this point, the trigonometric or single-sided DFT has been used. This pair is given by

$$x_n(s) = X_n(0) + \sum_{k=1}^{K-1}\left[\cos(2\pi ks/S)\quad \sin(2\pi ks/S)\right]\begin{bmatrix}X_n^C(k)\\X_n^S(k)\end{bmatrix}$$

$$\tag{6.3}$$

$$X_n(k) = \begin{bmatrix}X_n^C(k)\\X_n^S(k)\end{bmatrix} = \frac{2 - \delta(k)}{S}\sum_{s=0}^{S-1}\begin{bmatrix}\cos(2\pi ks/S)\\\sin(2\pi ks/S)\end{bmatrix}x_n(s) \tag{6.4}$$

where s is the time index, $S = 2K - 1$, and δ is the Kronecker delta function, which is defined as

$$\delta(k) = \begin{cases} 1 & \text{if } k = 0 \\ 0 & \text{if } k \neq 0 \end{cases}.$$

The second common DFT pair is the exponential or double-sided DFT.

$$x_n(s) = \sum_{k=1-K}^{K-1} X_n(k)e^{j2\pi ks/S} \qquad (6.5)$$

$$X_n(k) = \frac{1}{S}\sum_{s=0}^{S-1} x_n(s)e^{-j2\pi ks/S} \qquad (6.6)$$

where $X_n(k) = X_n^R(k) + jX_n^I(k)$. The single- and double-sided coefficients are related by

$$X_n^C(k) = \begin{cases} X_n^R(k) & \text{if } k = 0 \\ 2X_n^R(k) & \text{if } k \neq 0 \end{cases} \qquad X_n^S(k) = \begin{cases} -X_n^I(k) & \text{if } k = 0 \\ -2X_n^I(k) & \text{if } k \neq 0 \end{cases}$$

where $X_n^S(0) = X_n^I(0) = 0$ and the C, S, R, and I superscripts are used to denote the cosine, sine, real, and imaginary coefficients.

In order to simplify the notation used in the following derivation, the trigonometric DFT pair is used with the exponential coefficients. The resulting DFT pair is

$$x_n(s) = X_n^R(0) + 2\sum_{k=1}^{K-1}\left[\cos(2\pi ks/S) \quad -\sin(2\pi ks/S)\right]\begin{bmatrix} X_n^R(k) \\ X_n^I(k) \end{bmatrix}$$

$$(6.7)$$

$$X_n(k) = \begin{bmatrix} X_n^R(k) \\ X_n^I(k) \end{bmatrix} = \frac{1}{S}\sum_{s=0}^{S-1}\begin{bmatrix} \cos(2\pi ks/S) \\ -\sin(2\pi ks/S) \end{bmatrix}x_n(s) \qquad (6.8)$$

The harmonic balance equation (5.5) still holds with v related to V by (6.7) and I and Q related to i and q by (6.8). Applying Newton-Raphson to solve this equation results in the iteration

$$J_F(V^{(j)})(V^{(j+1)} - V^{(j)}) = -F(V^{(j)}) \tag{6.9}$$

where

$$J_F(V) = \left[J_{F,mn}(V)\right] = \left[\frac{\partial F_m(V)}{\partial V_n}\right], \tag{6.10}$$

$$J_{F,mn}(V) = \left[J_{F,mn}(V,k,l)\right] = \left[\frac{\partial F_m(V,k)}{\partial V_n(l)}\right], \tag{6.11}$$

with $m, n \in \{1, 2, \ldots, N\}$; $k, l \in \{0, 1, \ldots, K-1\}$; and

$$J_{F,mn}(V,k,l) = \frac{\partial F_m(V,k)}{\partial V_n(l)} = \begin{bmatrix} \dfrac{\partial F_m^R(V,k)}{\partial V_n^R(l)} & \dfrac{\partial F_m^R(V,k)}{\partial V_n^I(l)} \\[3mm] \dfrac{\partial F_m^I(V,k)}{\partial V_n^R(l)} & \dfrac{\partial F_m^I(V,k)}{\partial V_n^I(l)} \end{bmatrix}. \tag{6.12}$$

This derivative consists of the sum of terms

$$\frac{\partial F_m(V,k)}{\partial V_n(l)} = \frac{\partial I_m(V,k)}{\partial V_n(l)} + \begin{bmatrix} 0 & -\omega_k \\ \omega_k & 0 \end{bmatrix} \frac{\partial Q_m(V,k)}{\partial V_n(l)} + Y_{mn}(k,l). \tag{6.13}$$

where

$$Y_{mn}(k,l) = \begin{bmatrix} Y_{mn}^R(k,l) & -Y_{mn}^I(k,l) \\ Y_{mn}^I(k,l) & Y_{mn}^R(k,l) \end{bmatrix}. \tag{6.14}$$

Only $\dfrac{\partial I_m(V,k)}{\partial V_n(l)}$ is derived, the derivation of $\dfrac{\partial Q_m(V,k)}{\partial V_n(l)}$ is similar.

Compute $I_m(V,k)$ using (6.8),

$$I_m(V,k) = \frac{1}{S}\sum_{s=0}^{S-1}\begin{bmatrix} \cos(2\pi ks/S) \\ -\sin(2\pi ks/S) \end{bmatrix} i_m(v(s)) \tag{6.15}$$

The waveform v is considered to be an implicit function of its spectrum V; and so the chain rule is employed to compute the derivative.

$$\frac{\partial I_m(V,k)}{\partial V_n(l)} = \frac{1}{S}\sum_{s=0}^{S-1}\begin{bmatrix}\cos(2\pi ks/S)\\-\sin(2\pi ks/S)\end{bmatrix}\frac{\partial i_m(v(s))}{\partial v_n(s)}\frac{\partial v_n(s)}{\partial V_n(l)} \tag{6.16}$$

Now the derivative of $v_n(s)$ is calculated using (6.7).

$$v_n(s) = V_n(0) + 2\sum_{k=1}^{K-1}\begin{bmatrix}\cos(2\pi ks/S) & -\sin(2\pi ks/S)\end{bmatrix}\begin{bmatrix}V_n^R(k)\\V_n^I(k)\end{bmatrix} \tag{6.17}$$

$$\frac{\partial v_n(s)}{\partial V_n(l)} = \begin{bmatrix}\frac{\partial v_n(s)}{\partial V_n^R(l)} & \frac{\partial v_n(s)}{\partial V_n^I(l)}\end{bmatrix} = (2-\delta(l))\begin{bmatrix}\cos(2\pi ls/S) & -\sin(2\pi ls/S)\end{bmatrix} \tag{6.18}$$

This derivative is substituted into (6.16).

$$\frac{\partial I_m(V,k)}{\partial V_n(l)} =$$

$$\frac{2-\delta(l)}{S}\sum_{s=0}^{S-1}\begin{bmatrix}\cos(\frac{2\pi ks}{S})\\-\sin(\frac{2\pi ks}{S})\end{bmatrix}\frac{\partial i_m(v(s))}{\partial v_n(s)}\begin{bmatrix}\cos(\frac{2\pi ls}{S}) & -\sin(\frac{2\pi ls}{S})\end{bmatrix} \tag{6.19}$$

$$= \frac{2-\delta(l)}{S}\sum_{s=0}^{S-1}\frac{\partial i_m(v(s))}{\partial v_n(s)}\Psi \tag{6.20}$$

where

$$\Psi = \begin{bmatrix}\cos(2\pi ks/S)\cos(2\pi ls/S) & -\cos(2\pi ks/S)\sin(2\pi ls/S)\\-\sin(2\pi ks/S)\cos(2\pi ls/S) & \sin(2\pi ks/S)\sin(2\pi ls/S)\end{bmatrix}$$

or

$$\Psi = \tfrac{1}{2}\begin{bmatrix} \psi_{11} & \psi_{12} \\ \psi_{21} & \psi_{22} \end{bmatrix}$$

with

$$\psi_{11} = \cos(2\pi(k+l)s/S) + \cos(2\pi(k-l)s/S)$$

$$\psi_{21} = -\sin(2\pi(k+l)s/S) - \sin(2\pi(k-l)s/S)$$

$$\psi_{12} = -\sin(2\pi(k+l)s/S) + \sin(2\pi(k-l)s/S)$$

$$\psi_{22} = \cos(2\pi(k-l)s/S) - \cos(2\pi(k+l)s/S)$$

Define $G_{mn}(k) \in \mathbf{C}$ as the kth harmonic of $\dfrac{\partial i_m(v(s))}{\partial v_n(s)}$, i.e., let

$$G_{mn}(V,k) = \begin{bmatrix} G^R_{mn}(V,k) \\ G^I_{mn}(V,k) \end{bmatrix} = \frac{1}{S}\sum_{s=0}^{S-1} \frac{\partial i_m(v(s))}{\partial v_n(s)} \begin{bmatrix} \cos(2\pi ks/S) \\ -\sin(2\pi ks/S) \end{bmatrix}.$$

$$(6.21)$$

Then

$$\frac{\partial I_m(V,k)}{\partial V_n(l)} =$$

$$\frac{2-\delta(l)}{2}\begin{bmatrix} G^R_{mn}(k+l) + G^R_{mn}(k-l) & G^I_{mn}(k+l) - G^I_{mn}(k-l) \\ G^I_{mn}(k+l) + G^I_{mn}(k-l) & G^R_{mn}(k-l) - G^R_{mn}(k+l) \end{bmatrix}.$$

$$(6.22)$$

Equation (6.22) shows that the portion of the conversion matrices due to the nonlinear resistors $(\partial I_m(V)/\partial V_n)$ and capacitors $(\partial Q_m(V)/\partial V_n)$ can be split into the sum of a Toeplitz and a Hankel matrix[2]. For example

---

[2] A Toeplitz matrix has the form given by $a_{ij} = t_{i-j}$ and similarly, the form of a Hankel matrix is given by $a_{ij} = h_{i+j}$

$$\frac{\partial I_m(V)}{\partial V_n} = (T + H)D$$

where

$$T = \begin{bmatrix} t_0 & t_{-1} & t_{-2} & t_{-3} \\ t_1 & t_0 & t_{-1} & t_{-2} \\ t_2 & t_1 & t_0 & t_{-1} \\ t_3 & t_2 & t_1 & t_0 \end{bmatrix}$$

$$H = \begin{bmatrix} h_0 & h_1 & h_2 & h_3 \\ h_1 & h_2 & h_3 & h_4 \\ h_2 & h_3 & h_4 & h_5 \\ h_3 & h_4 & h_5 & h_6 \end{bmatrix}$$

$$D = \begin{bmatrix} \frac{1}{2} & & & \\ & 1 & & \\ & & 1 & \\ & & & 1 \end{bmatrix}$$

From (6.22),

$$t_k = \begin{bmatrix} G_{mn}^R(k) & -G_{mn}^I(k) \\ G_{mn}^I(k) & G_{mn}^R(k) \end{bmatrix} \qquad h_k = \begin{bmatrix} G_{mn}^R(k) & G_{mn}^I(k) \\ G_{mn}^I(k) & -G_{mn}^R(k) \end{bmatrix}.$$

It is the nature of the DFT that when applied to real waveforms $G_{nm}(k) = G_{nm}^*(-k)$ when $k < 0$ and $G_{nm}(k) = G_{nm}^*(2K-1-k)$ when $k > K$, where $*$ represents the complex conjugate operation in $\mathbf{C}$ [brigham74]. As a result, the Toeplitz and Hankel portions are rewritten

$$T = \begin{bmatrix} t_0 & t_1^* & t_2^* & t_3^* \\ t_1 & t_0 & t_1^* & t_2^* \\ t_2 & t_1 & t_0 & t_1^* \\ t_3 & t_2 & t_1 & t_0 \end{bmatrix} \qquad H = \begin{bmatrix} h_0 & h_1 & h_2 & h_3 \\ h_1 & h_2 & h_3 & h_3^* \\ h_2 & h_3 & h_3^* & h_2^* \\ h_3 & h_3^* & h_2^* & h_1^* \end{bmatrix}$$

This completes the derivation of the harmonic Jacobian $J_F(V)$. It is the synthesis of equations (6.10-6.22). This derivation allows the conversion matrices to be constructed with one FFT and $4K^2$ additions, considerably fewer operations than the $8K^3$ multiplications and additions needed for the previous derivation.

For a one node circuit at three frequencies the complete harmonic Jacobian would be

$$J_F(V) = \frac{\partial I(V)}{\partial V} + \Omega \frac{\partial Q(V)}{\partial V} + Y$$

where

$$\frac{\partial I(V)}{\partial V} = (T + H)D$$

where

$$T = \begin{bmatrix} G^R(0) & 0 & G^R(1) & G^I(1) & G^R(2) & G^I(2) \\ 0 & G^R(0) & -G^I(1) & G^R(1) & -G^I(2) & G^R(2) \\ G^R(1) & -G^I(1) & G^R(0) & 0 & G^R(1) & G^I(1) \\ G^I(1) & G^R(1) & 0 & G^R(0) & -G^I(1) & G^R(1) \\ G^R(2) & -G^I(2) & G^R(1) & -G^I(1) & G^R(0) & 0 \\ G^I(2) & G^R(2) & G^I(1) & G^R(1) & 0 & G^R(0) \end{bmatrix}$$

$$H = \begin{bmatrix} G^R(0) & 0 & G^R(1) & G^I(1) & G^R(2) & G^I(2) \\ 0 & -G^R(0) & G^I(1) & -G^R(1) & G^I(2) & -G^R(2) \\ G^R(1) & G^I(1) & G^R(2) & G^I(2) & G^R(2) & -G^I(2) \\ G^I(1) & -G^R(1) & G^I(2) & -G^R(2) & -G^I(2) & -G^R(2) \\ G^R(2) & G^I(2) & G^R(2) & -G^I(2) & G^R(1) & -G^I(1) \\ G^I(2) & -G^R(2) & -G^I(2) & -G^R(2) & -G^I(1) & -G^R(1) \end{bmatrix}$$

$$D = \begin{bmatrix} \tfrac{1}{2} & & & \\ & 1 & & \\ & & 1 & \\ & & & 1 \end{bmatrix}$$

$$\frac{\partial I(V)}{\partial V} = \begin{bmatrix} G^R(0) & 0 & 2G^R(-1) & 2G^I(1) \\ 0 & 0 & 0 & 0 \\ G^R(1) & 0 & G^R(0)+G^R(2) & G^I(2) \\ G^I(1) & 0 & G^I(2) & G^R(0)-G^R(2) \\ G^R(2) & 0 & G^R(1)+G^R(2) & -G^I(1)-G^I(2) \\ G^I(2) & 0 & G^I(1)-G^I(2) & G^R(1)-G^R(2) \end{bmatrix}$$

$$\begin{bmatrix} 2G^R(2) & 2G^I(2) \\ 0 & 0 \\ G^R(1)+G^R(2) & G^I(1)+G^I(2) \\ -G^I(1)-G^I(2) & G^R(1)-G^R(2) \\ G^R(0)+G^R(1) & -G^I(1) \\ -G^I(1) & G^R(0)-G^R(1) \end{bmatrix}$$

$\dfrac{\partial Q(V)}{\partial V}$ is similar with G replaced by $C = \mathbb{F}\dfrac{\partial q(v(s))}{\partial v(s)}$.

$$Y = \begin{bmatrix} Y^R(0) & 0 & 0 & 0 & 0 & 0 \\ 0 & 0 & 0 & 0 & 0 & 0 \\ 0 & 0 & Y^R(1) & -Y^I(1) & 0 & 0 \\ 0 & 0 & Y^I(1) & Y^R(1) & 0 & 0 \\ 0 & 0 & 0 & 0 & Y^R(2) & -Y^I(2) \\ 0 & 0 & 0 & 0 & Y^I(2) & Y^R(2) \end{bmatrix}$$

$$\Omega = \begin{bmatrix} 0 & 0 & 0 & 0 & 0 & 0 \\ 0 & 0 & 0 & 0 & 0 & 0 \\ 0 & 0 & 0 & -\omega_o & 0 & 0 \\ 0 & 0 & \omega_o & 0 & 0 & 0 \\ 0 & 0 & 0 & 0 & 0 & -2\omega_o \\ 0 & 0 & 0 & 0 & 2\omega_o & 0 \end{bmatrix}$$

Note that the second row and column of these conversion matrices consist completely of zeros, an artifact that results because phasors at DC must be real. This structural singularity in J_F can be removed either by deleting the offending row and column or making the diagonal entry nonzero and always setting the DC imaginary term to zero in the right-hand-side vector.

If not all frequencies computed in the transform are used in the harmonic balance calculations, then the resulting conversion matrix should be constructed as if all frequencies were included, except the rows and columns that correspond to the missing frequencies are deleted. This situation occurs if the FFT were used and required more frequencies than the user requested or if quasiperiodic signals were being transformed and the truncation used left some holes in the resulting translated set of frequencies.

3.1. Harmonic Relaxation-Newton

The second way to improve the harmonic Newton algorithm is to exploit the structure of the Jacobian to reduce the time required to factor it. Factoring the Jacobian is the most expensive operation required in the harmonic Newton algorithm. The techniques presented in this section are designed to reduce the expense of factoring this matrix. As a side benefit, they also speed its construction and the process of forward and backward substitution.

The Jacobian is organized as a block node admittance matrix that is sparse. Conventional sparse matrix techniques can be used to exploit its sparsity [duff86]. Each block is a conversion matrix that is itself a block matrix, consisting of 2×2 blocks that result from Fourier coefficients being members of **C**. Conversion matrices are full if they are associated with a node that has a nonlinear device attached, otherwise they are diagonal. In an integrated circuit, nonlinear devices attach to most nodes, so the conversion matrices will in general be full. It often happens, though, that nonlinear devices are either not active or are behaving very linearly. For example, the base-collector junction of a bipolar transistor that is in the forward-active region is reverse biased, and so the junction contributes nothing to its conversion matrices. If there are no other contributions to those conversion matrices, they may be ignored. If there are only contributions from linear components, they are diagonal. During the factorization, it is desirable to keep track of which conversion matrices are full, which are diagonal, and which are zero, and avoid unnecessary operations on known zero conversion matrix elements.

If a circuit contains many linear components, such as is common with hybrid microwave circuits, then much of the matrix will consist of diagonal blocks that are constant from iteration to iteration. It is possible to order the Jacobian during factorization such that the rows and columns associated with nodes that are attached only to linear devices are placed in the upper left corner of the matrix using variability types [norin71] [hachtel72]. These rows and columns are eliminated once before the Newton iteration starts. The elimination proceeds until a full block is encountered. In this way, the harmonic Jacobian is reduced in size by this prefactorization step until the number of rows and columns is equal to, or slightly greater than, the number of nodes with nonlinear devices attached.

A common approach to harmonic balance is to separate the linear devices into their own subcircuit and evaluate them once. A y-parameter matrix is created that describes the linear subcircuit to the harmonic balance analysis, this matrix becomes Y in (5.5). In this way, N is reduced from the number of nodes in the circuit to the number of nodes with nonlinear devices attached. The linear subcircuit approach is similar to using variability types, with one important difference. The variability types approach can be integrated into the normal pivot selection algorithm for the sparse matrix solver. The pivot selection algorithm, now fortified with variability types, can intelligently choose which of the nodes to eliminate and in which order. The order of elimination plays an important role in determining how many operations required for the factorization of a sparse matrix. It is possible to constrain the order in which the rows and columns are eliminated and end up with a reduced matrix that is identical to the one that results from the linear subcircuit approach. However, with variability types, the order is not constrained and it is likely that a better ordering will be found. Thus, using variability types is at least as efficient as using the linear subcircuit approach.

3.1.1. Adaptively Pruning the Jacobian

Applying traditional sparse matrix techniques is not enough to solve the Newton-Raphson iteration (6.9) efficiently. It is also necessary to reduce the density of the matrix. The Jacobian is only used to generate new iterates; it is not used when confirming convergence, so errors in the Jacobian only affect the rate and region of convergence, not the accuracy of the final solution. Approximations in the Jacobian reduce the asymptotic rate of convergence, but the gain in efficiency can more than make up for this loss. One approximation of this sort is Šamanskii's method, presented in Chapter 3, which results in a dramatic speedup for most circuits.

Another approach to approximating the Jacobian and thus speeding the iteration, results from exploiting the natural characteristics of conversion matrices for the nonlinear devices. As mentioned previously, these matrices are the sum of a Toeplitz and a Hankel matrix. Recall that

$$\frac{\partial I_m(V)}{\partial V_n} = (T + H)D$$

where

$$T = \begin{bmatrix} t_0 & t_{-1} & t_{-2} & t_{-3} \\ t_1 & t_0 & t_{-1} & t_{-2} \\ t_2 & t_1 & t_0 & t_{-1} \\ t_3 & t_2 & t_1 & t_0 \end{bmatrix}$$

$$H = \begin{bmatrix} h_0 & h_1 & h_2 & h_3 \\ h_1 & h_2 & h_3 & h_4 \\ h_2 & h_3 & h_4 & h_5 \\ h_3 & h_4 & h_5 & h_6 \end{bmatrix}$$

$$D = \begin{bmatrix} \frac{1}{2} & & & \\ & 1 & & \\ & & 1 & \\ & & & 1 \end{bmatrix}$$

$$t_k = \begin{bmatrix} G_{mn}^R(k) & -G_{mn}^I(k) \\ G_{mn}^I(k) & G_{mn}^R(k) \end{bmatrix} \qquad h_k = \begin{bmatrix} G_{mn}^R(k) & G_{mn}^I(k) \\ G_{mn}^I(k) & -G_{mn}^R(k) \end{bmatrix}$$

where G_{mn} is the sum of the derivative spectra for nonlinear resistors between node m and n, or from m to ground if $m = n$. This spectrum has the characteristic that the more linear the devices that generate it are behaving, the more the DC component dominates over the harmonics and the faster their magnitude drops off at higher harmonics. As a result, elements in the conversion matrix far from the diagonal will be small compared to those on the diagonal. To reduce the density of the harmonic Jacobian, these small terms far from the diagonal will be ignored.

Definition: The *guard harmonic* for a derivative spectrum is the smallest harmonic k such that

$$|G(l)| < \mu |G(0)| \quad \text{for all } l \geq k$$

where μ is a threshold parameter. (Typically $\mu = 10^{-4}$.)

When constructing the conversion matrices, (i.e., the blocks in the harmonic Jacobian resulting from nonlinear devices) all harmonics in the derivative spectrum used to form the conversion matrix are considered negligible if they are above the guard harmonic. These harmonics are set to zero, making the conversion matrices banded about the diagonal with the bandwidth an increasing function of how nonlinear the devices contributing to the matrix are behaving. Note that if the bandwidth is restricted to one, so all entries off the main diagonal of a conversion matrix are set to zero, then harmonic Newton collapses to block Gauss-Jacobi-Newton harmonic relaxation.

Ignoring those harmonics of the derivative spectra that fall above the guard harmonics greatly increases the initial sparsity of the harmonic Jacobian, however the Jacobian tends to fill-in while factoring it into L and U. To see this, consider the 3×3 banded block matrix in Figure 6.3. The original nonzeros are marked with crosses ($\times$) and the fill-ins are marked with circles (O). Notice the tendency of the bandwidth to increase in the blocks remaining after a major row and column have been eliminated. Also notice that, of the original nonzeros, those furthest from the diagonal of a block are due to the guard harmonics. These elements are small compared to the diagonal. The fill-ins inside the blocks always involve the guard harmonic, and so these fill-ins are assumed to be negligible. This heuristic does not have a sound theoretical basis, but is usually true if both the blocks

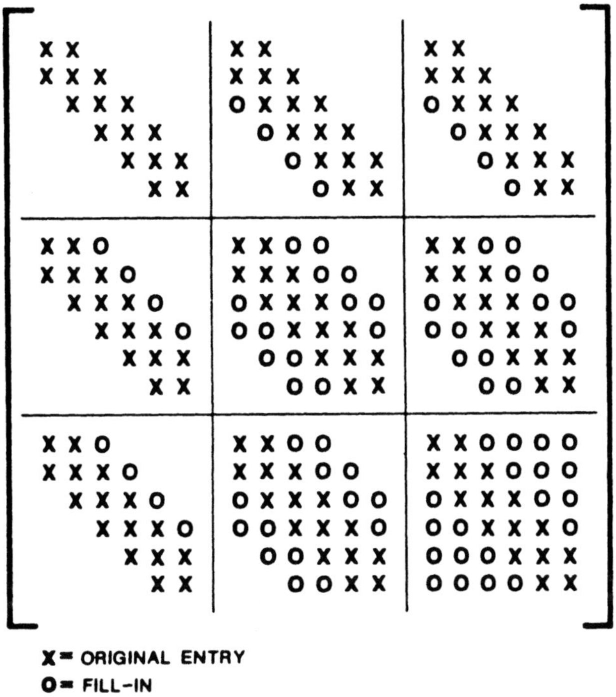

```
⎡  X X          X X          X X        ⎤
⎢  X X X        X X X        X X X      ⎥
⎢    X X X      O X X X      O X X X    ⎥
⎢      X X X      O X X X      O X X X  ⎥
⎢        X X X      O X X X      O X X X⎥
⎢          X X        O X X        O X X⎥
⎢                                       ⎥
⎢  X X O        X X O O      X X O O    ⎥
⎢  X X X O      X X X O O    X X X O O  ⎥
⎢    X X X O    O X X X O O  O X X X O O⎥
⎢      X X X O  O O X X X O  O O X X X O⎥
⎢        X X X    O O X X X    O O X X X⎥
⎢          X X      O O X X      O O X X⎥
⎢                                       ⎥
⎢  X X O        X X O O      X X O O O O⎥
⎢  X X X O      X X X O O    X X X O O O⎥
⎢    X X X O    O X X X O O  O X X X O O⎥
⎢      X X X O  O O X X X O  O O X X X O⎥
⎢        X X X    O O X X X    O O O X X X⎥
⎣          X X      O O X X    O O O O X X⎦
```

X = ORIGINAL ENTRY
O = FILL-IN

Figure 6.3: Fill-in pattern of a banded block matrix.

and the block matrix are strongly diagonally dominant. Thus the non-linearities should be resistive and behaving only mildly nonlinear, and each node in the circuit should be connected to ground with an admittance that is large compared to the admittances connecting it to other nodes. These conditions are very restrictive and rarely satisfied in practice, however the heuristic works quite well if μ is small and usually results in a considerable speed up.

3.1.2. Harmonic Jacobian Data Structure

Implementing a task such as factoring the harmonic Jacobian on a computer not only requires a good algorithm, but also good data structures. The data structure used by *Spectre* to hold the harmonic Jacobian is a variation on the standard orthogonal-linked list used to hold sparse matrices [kundert86a]. The matrix is written as a block node admittance matrix. Thus, the matrix has N rows and columns, where N is the number of nodes in the circuit. Each element in the matrix contains a $2K \times 2K$ block, where K is the number of frequencies, and has a pointer to the element below and to the right of itself. The element also carries the bandwidth of the block. The block is allocated as a full $2K \times 2K$ matrix, but is treated numerically as a banded matrix with the given bandwidth. This approach allows the bandwidth to be set as needed on each iteration, allowing harmonic relaxation-Newton to adapt to the problem being solved.

4. Spectre

Spectre implements harmonic relaxation-Newton for periodic and quasiperiodic nonautonomous circuits [kundert]. With periodic signals, the FFT is used as the Fourier transform and the number of frequencies used is constrained to be a power of two. With quasiperiodic signals, the APFT is used. A version has been written that also implements the FFT for quasiperiodic circuits and one that implements periodic autonomous circuits. *Spectre* takes as input a file with a SPICE-like description of the circuit, that is a list of components (transistors, resistors, capacitors, transmission lines, etc) with their node connections, and a list of analyses to be performed. Along with harmonic balance, *Spectre* also performs conventional DC, AC, and S-parameter analyses. When a harmonic balance analysis is requested, the desired number of harmonics in the solution must be specified.

4.1. Comparisons to Transient Analysis

Spectre was used to simulate several circuits to exhibit some of the capabilities of the simulator and to contrast the performance of *Spectre* against the performance of a representative transient analysis simulator such as SPICE [nagel75]. The times for three circuits are presented in this section. The first two circuits are well suited to frequency-domain simulation and poorly suited to time-domain simulation. With the third circuit, the situation is reversed.

The first circuit is the traveling-wave amplifier (TWA) shown in Figure 6.4. It contains four GaAs MESFETs and ten transmission lines of noncommensurate length. While *Spectre* naturally handles lossy and dispersive lines, the lines were constrained to be ideal to be compatible with SPICE. The run times for SPICE2, SPICE3, and *Spectre* are shown in Table 6.1. The long simulation time required by SPICE2 results from the particular break-point algorithm used in this program. This algorithm is very inefficient for circuits containing transmission lines. While the problem is not inherent to transient analysis, the wide spread use of SPICE2 has created the misconception that transient analysis is not suitable for microwave circuits. As a partial solution, SPICE3 [quarles89] allows the break-point algorithm to ignore the transmission lines resulting in a shorter simulation time at the expense of increased risk of incorrect answers.

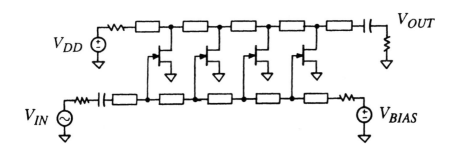

Figure 6.4: A four-segment traveling-wave amplifier.

Spectre does not yet have automatic error control algorithms, so the user is forced to specify the number of harmonics. This circuit was simulated with 8, 16, and 32 harmonics. On this particular circuit, 8 harmonics resulted in accuracy commensurate with SPICE. Note that doubling the number of harmonics more than doubled the time required to complete the simulation, but the increase is well below the factor of 8 that would be expected from straight harmonic Newton.

Figure 6.5 shows the results computed by *Spectre* for the traveling-wave amplifier when the power of the input source was swept from -20 dBm to 20 dBm in 40 steps. Notice that at the highest input power, the amplifier exhibited 10 dB of compression. The time required to perform this analysis is shown in Table 6.2. Swept analyses in harmonic balance exploits the useful characteristic of the Newton-Raphson algorithm that it converges faster and is more likely to converge if primed with an initial guess close to the solution. With a swept analysis, *Spectre* calculates an initial guess by extrapolating from the solution at the previous step. Thus, a sweep with 40 points is considerable faster than 40 individual analyses. This feature is an important advantage harmonic balance has over transient analysis.

TWA Steady State		
SPICE2		62500
SPICE3		240
Spectre	8 harmonics	7
	16 harmonics	22
	32 harmonics	56

Table 6.1: Time required to simulate the traveling-wave amplifier of Figure 6.4. The simulation interval for SPICE was two periods of the input. The period of the input signal was 10 ns while the electrical lengths of the various transmission lines were 86 ps, 187 ps, and 251 ps. Times were measured on a VAX785.

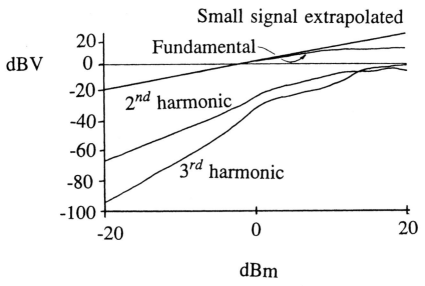

Figure 6.5: The power level of the first three harmonics of the output of the traveling-wave amplifier of Figure 6.4 versus input power.

TWA Power Sweep		
Spectre	4 harmonics	38
	8 harmonics	130
	16 harmonics	800

Table 6.2: Time required to simulate the traveling-wave amplifier of Figure 6.4 over the input power levels shown in Figure 6.5 in 40 steps. Times were measured on an HP9000/350 with floating point accelerator.

Spectre computed the intermodulation distortion of the traveling-wave amplifier by applying two 200 mV signals, one each at 10 GHz and 10.4 GHz. The response is shown, both in the time and frequency domains, in Figure 6.6. The circuit was simulated with $H = 5$ using the diamond truncation given in (3.6). The computation time and memory requirements for several values of H and for both box (3.5) and diamond (3.6) truncation schemes are shown in Table 6.3. There are a few comments that should be made to clarify some of the results in the table. *Spectre*'s memory allocator expands array sizes in factors of two, which is why memory requirements sometimes do not change even though H changes. Each doubling of the array size quadruples the amount of memory required. Most of the approximate factor of two differences between physical and virtual memory requirements can be eliminated by better implementation. Any simulation that needed over 64 frequencies required more memory than the 44 megabytes available from the operating system.

The frequencies of the two tones were chosen so that the various spectral lines in Figure 6.6 could be resolved by eye. The circuit was resimulated with the frequency of the tones set as close as 10 GHz and 10GHz + 1 Hz with no apparent change of accuracy or efficiency. The 1 Hz separation in the two tones results in an intermodulation product at 1 Hz. The combination of 10 GHz and 1 Hz signals make it prohibitively expensive to find the steady-state response of this circuit with a transient analysis simulator such as SPICE.

Times required to simulate the self-biasing FET tuned amplifier shown in Figure 2.1 is compared in Table 6.4. This circuit is troublesome to transient analysis because of the slow time constant of the input DC blocking capacitor. Normally, time constants due to bias circuits are avoided by using the DC solution as the initial condition. With this circuit, the DC gate voltage is affected by the large input signals. It is very difficult to predict the settling time, and therefore the required simulation time, because of the nonlinear effect of the gate diode on the bias circuit time constant. The simulation interval was chosen for this circuit by simulating the circuit several times. Each time the simulation interval was increased and the result compared with

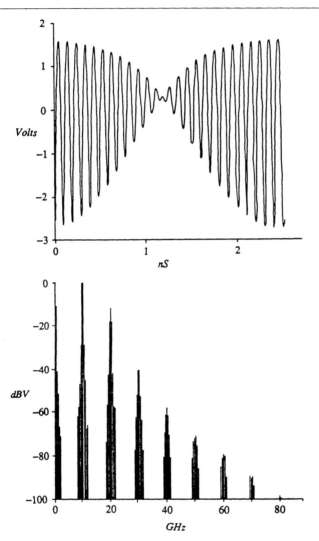

Figure 6.6: Response of traveling-wave amplifier of Figure 6.4 to two-tone input.

TWA Intermodulation Distortion				
H	K	Time	Physical Memory	Virtual Memory
Using Λ_K generated by (3.5).				
1	5	0.63 s	0.55 MB	0.78 MB
2	13	4.2 s	0.87 MB	1.5 MB
3	25	24 s	2.2 MB	3.9 MB
4	41	98 s	7.5 MB	14 MB
5	61	320 s	7.6 MB	14 MB
Using Λ_K generated by (3.6).				
1	3	0.35 s	0.50 MB	0.80 MB
2	7	1.3 s	0.50 MB	0.80 MB
3	13	4.4 s	0.87 MB	1.5 MB
4	21	15.6 s	2.2 MB	3.9 MB
5	31	43 s	2.3 MB	4.0 MB
6	43	110 s	7.5 MB	14 MB
7	57	245 s	7.6 MB	14 MB

Table 6.3: Execution times and memory requirements for *Spectre* running an intermodulation distortion test on the traveling-wave amplifier of Figure 6.4. Times were measured on a VAX8800.

the previous simulation until it became clear that the circuit had fully settled. The time reported was that required by SPICE to simulate the circuit over the shortest interval for which the circuit settled to within 0.1% of its steady-state value. *Spectre* found the steady-state solution of this circuit directly without forcing the user to know the circuit's settling time.

The last circuit, shown in Figures 6.7 and 6.8, is a simple noninverting amplifier based on a *uA* 741 op amp. Results for this circuit are presented in Table 6.5 with various output signal levels and load resistances. The circuit contains no long time constants and so transient analysis can be used to compute the steady-state response efficiently. With large output current, the op amp behaves internally

Self-Biased Amplifier		
SPICE2		608
Spectre	8 harmonics	1.4
	16 harmonics	5
	32 harmonics	28

Table 6.4: Time required to simulate the self-biased amplifier of Figure 2.1. Times were measured on a VAX785.

strongly nonlinear, which slows harmonic balance. Indeed, the output stage goes into class B operation at high output currents. This example shows that harmonic balance is able to handle strongly nonlinear circuits, though it may require more time than traditional transient analysis. It should be mentioned that at low input levels with no load resistor, harmonic balance gives a much more accurate answer and requires half the analysis time.

4.2. Profiles

With transient analysis, the time required to evaluate the nonlinear device equations generally dominates over all other tasks except on very large circuits, where the time required to factor and solve the large system of linear equations dominates. The transition point where the time required of both tasks is equal generally occurs between 1000 and 10,000 nodes, depending on the circuit and the complexity of the models. The situation is completely different with harmonic balance, where the time required for operations involving the system of linear equations completely dominates over all other tasks. For example, consider the circuit shown in Figure 6.9, a seven stage traveling-wave amplifier with 14 GaAsFETs and 24 transmission lines [orr86]. Table 6.6 shows the time required to evaluate device models (eval), convert signals into and out of the frequency domain (Fourier), and construct, factor, and solve the linear system of equations (eqns) for various number of frequencies and power levels. The time required for handling the linear system of equations was further resolved in Table 6.7

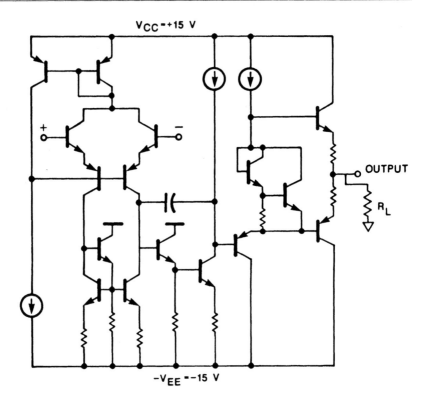

Figure 6.7: Schematic of *uA* 741 operational amplifier.

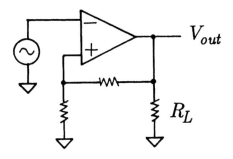

Figure 6.8: Op amp of Figure 6.7 shown in a noninverting amplifier configuration with gain of 100.

μa741 Op Amp				
			Spectre	
Conditions	*SPICE2*		Harmonics	
		8	16	32
$V_{out} = 1V$ $R_L = \infty$	9	5	11	25
$V_{out} = 1V$ $R_L = 10K$	13	8	19	40
$V_{out} = 10V$ $R_L = 10K$	14	32	132	497

Table 6.5: Time required to simulate the amplifier of Figure 6.8. Times were measured on a VAX785.

where the time required to construct, factor, and solve the factored system of equations is shown.

These results show that the most expensive aspect of harmonic balance is constructing, factoring, and solving the linear system of

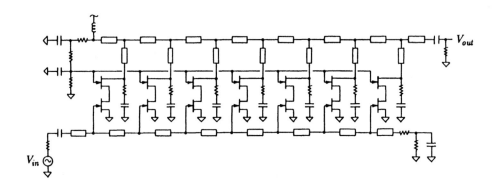

Figure 6.9: A GaAsFET traveling-wave amplifier [orr86].

Spectre Profiles							
H_1	H_2	K	P_{in} (dBm)	t_{total} (sec)	Eval (%)	Fourier (%)	Eqns (%)
4	0	4	-10	6.26	1.6	3.5	74.5
4	0	4	0	6.8	2.1	4.1	73.2
4	0	4	10	16.9	4.4	9.6	74.1
8	0	8	-10	15.5	0.5	3.2	86.9
8	0	8	0	26.38	0.7	1.9	91.5
8	0	8	10	68.32	2.3	4.8	89.5
16	0	16	-10	36.34	0.8	2.4	87.5
16	0	16	0	155.72	0.3	0.6	97.1
16	0	16	10	433.88	0.6	1.4	96.8
4	4	13	-10	186.24	0.2	1.0	97.2
4	4	13	0	267.34	0.1	1.0	96.6

Table 6.6: Time in seconds required for various portions of the *Spectre* code for the circuit of Figure 6.9 with various number of frequencies and power levels. Each run required only one full Newton iteration, but the runs at the higher power levels required many Šamanskii iterations (where the Jacobian was reused). H_i represents the number of harmonics for the ith fundamental, and K is the total number of frequencies. Times were measured on an HP9000/350 with floating point accelerator.

Spectre Matrix Profiles						
K	P_{in} (dBm)	t_{total} (sec)	Construct (%)	Factor (%)	Solve (%)	Iters (N/S)
4	-10	6.26	10.1	62.8	6.7	1/1
4	0	6.8	11.0	59.0	9.4	1/2
4	10	16.9	25.4	31.0	31.8	1/21
8	-10	15.5	7.0	77.0	6.6	1/1
8	0	26.38	5.4	82.3	6.4	1/2
8	10	68.32	13.4	55.7	27.5	1/22
16	-10	36.34	6.1	78.7	5.9	1/1
16	0	155.72	2.0	92.9	3.2	1/2
16	10	433.88	4.1	78.7	16.0	1/24
13	-10	186.24	4.9	90.7	2.8	1/1
13	0	267.34	4.2	89.6	3.9	1/3

Table 6.7: Time in seconds required for various portions of the *Spectre* matrix code for the circuit in Figure 6.9 with various number of frequencies and power levels. The time required to construct the matrix included the time to evaluate the device model equations and convert signals to and from the time domain (Construct). The remaining times are those required to factor and solve the linear system of equations. The number of full Newton / Šamanskii iterations are given. Times were measured on an HP9000/350 with floating point accelerator.

equations. This is true even with harmonic relaxation-Newton and Šamanskii's method being used.

4.3. Harmonic Relaxation vs. Harmonic Newton

This section explores how reducing the bandwidth of the conversion matrices affects the time required to complete a simulation. In particular, harmonic relaxation, harmonic relaxation-Newton and harmonic Newton are compared versus computation time and number of iterations. It is expected that harmonic relaxation be the fastest but least robust, harmonic Newton be the most robust but slowest, and that harmonic relaxation-Newton should be a robust method that adapts to the problem to provide the speed of harmonic relaxation when it does not jeopardize convergence. Table 6.8 shows computation time and Table 6.9 shows the number of iterations for *Spectre* simulating the circuit of Figure 6.10 for a periodic solution containing 8 harmonics.

These results show that harmonic relaxation-Newton as implemented in *Spectre* with a fixed guard threshold of typically 0.0001 does

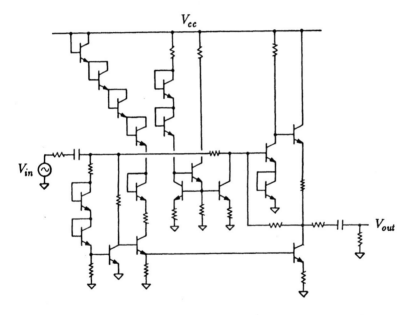

Figure 6.10: Class AB monolithic power amplifier[3].

---

[3]R. G. Meyer, private communication.

Spectre Computation Time vs. Method								
P_{in} (dBm)	HR	GJNHR	HRN					HN
			0.3	0.1	0.01	0.001	0.0001	
-20	∞	22	16	21	30	36	42	85
-15	∞	39	25	48	41	50	51	95
-10	∞	155	∞	63	77	88	101	136
-5	∞	∞	∞	∞	138	140	179	259
0	∞	∞	∞	∞	∞	413	439	518
5	∞	∞	∞	∞	∞	695	680	859
10	∞	∞	∞	∞	∞	1440	1420	1710

Table 6.8: Time in seconds required for *Spectre* to simulate the class A/B amplifier of Figure 6.10 using various harmonic balance methods. Methods include harmonic relaxation, both splitting (HR) and Gauss-Jacobi-Newton harmonic relaxation (GJNHR), harmonic relaxation newton (HRN) for various guard harmonic thresholds, and harmonic Newton (HN). Times were measured on an HP9000/350 with floating point accelerator.

			Spectre Iteration Count vs. Method					
P_{in} (dBm)	HR	GJNHR	HRN					HN
			0.3	0.1	0.01	0.001	0.0001	
-20	∞	1/9	1/4	1/4	1/5	1/5	1/5	1/5
-15	∞	1/18	1/7	2/9	1/8	1/8	1/8	1/8
-10	∞	3/78	∞	1/21	1/21	1/21	1/21	1/21
-5	∞	∞	∞	∞	3/19	3/18	3/18	3/18
0	∞	∞	∞	∞	∞	5/63	5/63	5/63
5	∞	∞	∞	∞	∞	10/60	10/60	10/60
10	∞	∞	∞	∞	∞	21/113	22/94	22/94

Table 6.9: Iterations required for *Spectre* to simulate the class A/B amplifier of Figure 6.10 using various harmonic balance methods. Methods include harmonic relaxation, both splitting (HR) and Gauss-Jacobi-Newton harmonic relaxation (GJNHR), harmonic relaxation newton (HRN) for various guard harmonic thresholds, and harmonic Newton (HN). The first number given is the number of full Newton iterations, the second is the number of Šamanskii iterations.

not extract as much of the speed of Gauss-Jacobi harmonic relaxation as is possible. Extracting that speed is more complicated than just changing the guard harmonic threshold, as shown in the following tables. Table 6.10 shows computation time and Table 6.11 shows the number of iterations for *Spectre* simulating the circuit of Figure 6.9 for a periodic solution containing 16 harmonics.

With the traveling-wave amplifier, for large thresholds (but less than one), convergence for harmonic relaxation-Newton is worse than for harmonic relaxation. It appears as if truncating the conversion matrices to be diagonal gives better convergence properties than does truncating the conversion matrices to have a small bandwidth greater than one. This odd behavior has yet to be explained.

Spectre Computation Time vs. Method								
P_{in} (dBm)	HR	GJNHR	HRN					HN
			0.3	0.1	0.01	0.001	0.0001	
-10	∞	3	3	6	12	15	14	119
-5	∞	3	3	6	11	14	17	123
0	∞	3	86	33	23	36	39	120
5	∞	5	76	186	118	102	100	136
10	∞	∞	428	144	153	162	147	185

Table 6.10: Time in seconds required for *Spectre* to simulate the traveling wave amplifier of Figure 6.9 using various harmonic balance methods with 16 harmonics. Methods include harmonic relaxation, both splitting (HR) and Gauss-Jacobi-Newton harmonic relaxation (GJNHR), harmonic relaxation newton (HRN) for various guard harmonic thresholds, and harmonic Newton (HN). Times were measured on an HP9000/350 with floating point accelerator.

Spectre Iteration Count vs. Method								
P_{in} (dBm)	HR	GJNHR	HRN					HN
			0.3	0.1	0.01	0.001	0.0001	
-10	∞	1/0	1/1	1/1	1/0	1/0	1/0	1/0
-5	∞	1/2	1/1	1/1	1/1	1/0	1/0	1/0
0	∞	1/3	4/7	1/1	1/1	1/1	1/1	1/1
5	∞	1/6	3/9	4/6	2/12	1/2	1/2	1/2
10	∞	∞	8/30	1/24	1/23	1/23	1/23	1/23

Table 6.11: Iterations required for *Spectre* to simulate the traveling wave amplifier of Figure 6.9 using various harmonic balance methods. Methods include harmonic relaxation, both splitting (HR) and Gauss-Jacobi-Newton harmonic relaxation (GJNHR), harmonic relaxation newton (HRN) for various guard harmonic thresholds, and harmonic Newton (HN). The first number given is the number of full Newton iterations, the second is the number of Šamanskii iterations. With a Šamanskii iteration, the Jacobian from the previous iteration is reused.

4.4. APFT versus FFT

In order to explore the advantage of implementing quasiperiodic harmonic balance using the FFT rather than the APFT, a two-tone intermodulation distortion test was performed on the traveling-wave amplifier of Figure 6.9. Two −10 dBm signals were applied to the amplifier at 1 GHz and 1.01 GHz and 3 harmonics of each signal were calculated along with sum and difference frequencies to a total of 13 frequencies. The time required on an HP9000/850 to simulate this circuit was 106 seconds for APFT-based harmonic Newton and 81 seconds for FFT-based harmonic relaxation-Newton[4]. The primary reason that the FFT-

---

[4]The FFT-based measurements were made using the HP85150B Microwave Nonlinear Simulator (MNS). This simulator is a direct descendant of *Spectre*, but has been upgraded to use the FFT rather than the APFT for its transform when performing quasiperiodic harmonic balance.

based method was faster was because it allowed the use of harmonic relaxation-Newton. When the circuit were resimulated with APFT-based harmonic relaxation-Newton, the time required was 82 seconds. Harmonic relaxation-Newton is not normally used with the APFT because the aliasing pattern of the APFT is incompatible with the guard harmonic heuristic. The speed up achieved on this circuit by using harmonic relaxation-Newton and the FFT is modest, considerably greater speedup can be achieved when much of the circuit is behaving near linearly or when a large number of frequencies are used.

Chapter 7
Mixed Frequency-Time Method

In this chapter a boundary constraint is developed that restricts the set of solutions of a differential equation to those that are quasiperiodic. A first attempt is made by using the periodic boundary constraint (3.21 c), but it is shown to be computationally too expensive. In the process it is discovered that there are periodic problems that are better handled with a quasiperiodic boundary constraint. Another approach was suggested by Chua and Ushida [chua81]. They construct an N-point quasiperiodic boundary constraint by, assuming that the quasiperiodic signals are accurately approximated by a Fourier series with just K frequencies $\Lambda_K = \{0, \omega_1, \ldots, \omega_{K-1}\}$, sampling the waveforms at $M > 2K - 1$ points $\tau = \{t_1, t_2, \ldots, t_M\}$, and insisting that the resulting sampled waveform is quasiperiodic (i.e., it belongs to $AP(\Lambda_K, \tau)$). This method trades off accuracy for efficiency, but the tradeoff is such that the method is impractical for almost all problems. A generalization of this approach, referred to as the mixed frequency-time method (MFT), avoids both the efficiency and accuracy problems of the previous methods and is discussed in much greater depth.

1. Previous Quasiperiodic Boundary Constraints

1.1. The Periodic Boundary Constraint

The periodic boundary constraint (3.21 c) can be used to find approximately quasiperiodic solutions to differential equations because an almost periodic function x always has the property that given some $\varepsilon > 0$ there always exists a $T \in \mathbb{R}$ such that

$$\| x(t+T) - x(t) \| < \varepsilon \qquad \text{for all} \ \ t \in \mathbb{R}. \qquad (7.1)$$

T is referred to as an ε-*almost period* of x. The collection of all ε-almost periods make up the ε-*translation set* of x [hale80].

   Given ε, it is very difficult to find an ε-almost period, however it is possible to find an almost period T for which $\max_{t} \| x(t+T) - x(t) \|$ is small, but not necessarily smaller than some ε specified in advance. One approach to finding such an almost period for $x \in QP(\lambda_1, \lambda_2, \ldots, \lambda_d; \mathbb{R})$ is to choose a frequency ν such that for each $j \in \{1, 2, \ldots, d\}$ there exists a k_j with $k_j \nu \approx \lambda_j$. The almost period is then $2\pi/\nu$. For example, consider $x(t) = \cos(t) + \cos(\pi t)$. This is 2-fundamental quasiperiodic with $\lambda_1 = 1$ and $\lambda_2 = \pi$. A reasonably good rational approximation to π is 355/113. So choose $\nu = 1/113$. Then $113\nu = \lambda_1$ and $355\nu = \lambda_2 + \delta$ where $|\delta| < 3 \times 10^{-7}$. Choosing $T = 2\pi\nu = 226\pi$ results in $|x(t+T) - x(t)| < 0.0002$ for all t.

   While it is always possible to find an almost period T that satisfies (7.1) with some ε sufficiently small, the resulting boundary-value problem is usually extraordinarily expensive to solve. For the example, the boundary constraint involves 355 periods of the largest fundamental. Since all methods that solve boundary-value problems require the differential equation to be integrated over the interval inclosed by the boundaries, using (7.1) as a quasiperiodic boundary constraint is usually prohibitively expensive. This problem also occurs when trying to find the periodic solution of a differential equation. Consider the example again, but this time let $\lambda_2 = 355/113$ from the beginning. The solution is now periodic, but the boundary constraint still involves 355 periods of fundamental λ_2. In this situation, it is best to use a method specifically developed for quasiperiodic problems, such as the mixed frequency-time method presented later in this chapter.

1.2. The N-Point Boundary Constraint

Consider the lumped quasiperiodic test problem (3.11). Recall that $v \in QP(\lambda_1, \lambda_2, \ldots, \lambda_d; \mathbb{R})$ and so

$$v(t) = V_0 +$$

$$\sum_{k_1=1}^{\infty} \cdots \sum_{k_d=1}^{\infty} \left[V^C(k_1, \ldots, k_d)\cos(sT(k_1\lambda_1 + \cdots + k_d\lambda_d)) + \right.$$

$$\left. V^S(k_1, \ldots, k_d)\sin(sT(k_1\lambda_1 + \cdots + k_d\lambda_d)) \right].$$

Assume that this series can be truncated to K frequencies without introducing significant error, and that the resulting set of frequencies is $\Lambda_K = \{0, \omega_1, \ldots, \omega_d\}$. If $S = 2K - 1$ distinct time-points are chosen and v is sampled at these points, then the S coefficients of the truncated Fourier series can be determined using the APFT. Once the coefficients are known, the series can be evaluated for the solution to the differential equation for any time t. If the assumption of only K frequencies being significant is correct, and if the signal being sampled is quasiperiodic, then the solution as computed by integrating the differential equation and the Fourier series should agree for all time. Thus, the quasiperiodic boundary constraint is formed by selecting another time-point (one that differs from those chosen previously) and insisting that the solution to the differential equation agree with the Fourier series.

For the lumped test problem (3.11)

$$i(v(t)) + \dot{q}(v(t)) + u(t) = 0. \tag{7.2}$$

Sample the solution at $S + 1$ time-points $\{t_1, t_2, \ldots, t_{S+1}\}$. Using Λ_K

and all but the last time-point, form the inverse APFT matrix Γ^{-1} (B.1) and solve for the Fourier coefficients

$$\Gamma^{-1}\begin{bmatrix} V_0 \\ V_1^C \\ V_1^S \\ \vdots \\ V_{K-1}^C \\ V_{K-1}^S \end{bmatrix} = \begin{bmatrix} v(t_1) \\ v(t_2) \\ v(t_3) \\ \vdots \\ v(t_S) \end{bmatrix}$$

Evaluate the Fourier series at t_{S+1}

$$\rho(t_{S+1})\begin{bmatrix} V_0 \\ V_1^C \\ V_1^S \\ \vdots \\ V_{K-1}^C \\ V_{K-1}^S \end{bmatrix} = v(t_{S+1})$$

where
$$\rho(t_{S+1}) = [1 \; \cos(\omega_1 t_{S+1}) \; \sin(\omega_1 t_{S+1}) \; \cdots \; \cos(\omega_{K-1} t_{S+1}) \; \sin(\omega_{K-1} t_{S+1})].$$
Thus the quasiperiodic boundary constraint on (7.2) is written in short form as

$$\rho(t_{S+1})\Gamma\begin{bmatrix} v(t_1) \\ v(t_2) \\ v(t_3) \\ \vdots \\ v(t_S) \end{bmatrix} - v(t_{S+1}) = 0 \tag{7.3}$$

For the calculation of Γ from Γ^{-1} to be well conditioned, the time-points should be distributed over at least one period of the smallest frequency in Λ_K (see Appendix B). This implies that the differential equation must be integrated for at least one period of the smallest frequency. As an example of how inefficient this can become, consider $v(t) = \sum_{k_1} \sum_{k_2} (\cos(k_1\lambda_1 t) + \cos(k_2\lambda_2 t))$ and truncate the number of frequencies by limiting $|k_1|, |k_2| \leq 3$. If $\lambda_1 = 1$ and $\lambda_2 = \pi$, as in the previous example, then $\omega_{min} = \pi - 3 \approx 0.1415$ and $\omega_{max} = 3\pi$. The ratio of ω_{max} to ω_{min} is 67, and so the differential equation must be evaluated for at least 67 periods of the highest frequency. This involves solving the differential equation over a much shorter interval than the that required when using the periodic boundary constraint, as we show in the next section, it is possible to reduce this interval further and remove the direct dependence of the length of the interval on the input frequencies.

2. The Mixed Frequency-Time Method

Devising a boundary constraint that restricts the set of possible solutions for a differential equation to those that are quasiperiodic involves blending together both time-domain and frequency-domain concepts. A practical approach for doing this is the Mixed Frequency-Time (or MFT) method [kundert88c, kundert89]. This method can be formulated as a two-point boundary constraint, but to do so obscures the fundamental ideas behind the method. The method is presented in the most natural manner first, and then it is shown that MFT converts a differential equation into a boundary-value problem.

The mixed frequency-time method begins with the assumption that all waveforms present are d-quasiperiodic and that the frequency of each of the d fundamentals is known. One fundamental, often the highest in frequency, is chosen to be the clock. The waveforms are then sampled at the beginning of each clock cycle as shown in Figure 7.1 where the beginning of the cycle is chosen arbitrarily, but must be consistent on each cycle. The resulting sampled waveforms are $(d-1)$-quasiperiodic, as stated in the following theorem.

Theorem 7.1 *(Quasiperiodic Sampling Theorem): Let* $x \in QP(\lambda_1, \lambda_2, \ldots, \lambda_d; \mathbb{R})$ *and* $T = 2\pi/\lambda_d$ *be the clock period. Let* $x_s = x(sT + \theta)$ *where* $s \in \mathbb{Z}$ *and* $\theta \in \mathbb{R}$ *be the clock phase. Then*

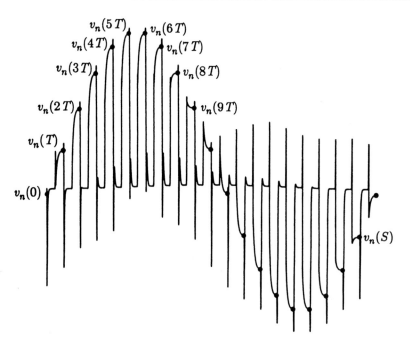

Figure 7.1: The response of a switching filter circuit to a periodic function, with the initial points of each cycle denoted.

$\{x_s\} \in QP(\lambda_1, \lambda_2, \ldots, \lambda_{d-1}; Z(T, \theta))$ *where* Z *is the set of sample times* $Z(T, \theta) = \{t : t = rT + \theta, r \in Z\}$.

Proof:

$$x(t) = \sum_{k_1=-\infty}^{\infty} \sum_{k_2=-\infty}^{\infty} \cdots \sum_{k_d=-\infty}^{\infty} X(k_1, k_2, \ldots, k_d) e^{j(k_1\lambda_1 + k_2\lambda_2 + \cdots + k_d\lambda_d)t}$$

$$x_s = x(2\pi s / \lambda_d + \theta)$$

$$x_s = \sum_{k_1=-\infty}^{\infty} \cdots \sum_{k_d=-\infty}^{\infty} X(k_1, k_2, \ldots, k_d) e^{\mathbf{j}(k_1\lambda_1 + \cdots + k_d\lambda_d)(2\pi s/\lambda_d + \theta)}$$

Let $\psi = (k_1\lambda_1 + k_2\lambda_2 + \cdots + k_{d-1}\lambda_{d-1})\theta$

$$x_s = \sum_{k_1=-\infty}^{\infty} \cdots \sum_{k_d=-\infty}^{\infty}$$

$$X(k_1, k_2, \ldots, k_d) e^{\mathbf{j}2\pi s\left[\frac{k_1\lambda_1}{\lambda_d} + \cdots + \frac{k_{d-1}\lambda_{d-1}}{\lambda_d} + k_d\right] + \mathbf{j}k_d\lambda_d\theta + \mathbf{j}\psi}$$

$$x_s = \sum_{k_1=-\infty}^{\infty} \cdots \sum_{k_d=-\infty}^{\infty}$$

$$X(k_1, k_2, \ldots, k_d) e^{\mathbf{j}2\pi s\left[\frac{k_1\lambda_1}{\lambda_d} + \cdots + \frac{k_{d-1}\lambda_{d-1}}{\lambda_d}\right]} e^{\mathbf{j}k_d(2\pi s + \lambda_d\theta)} e^{\mathbf{j}\psi}$$

Recall that $e^{\mathbf{j}2\pi s k_d} = 1$ for all s and k_d and so $e^{\mathbf{j}k_d(2\pi s + \lambda_d\theta)} = e^{\mathbf{j}k_d\lambda_d\theta}$.

$$x_s = \sum_{k_1=-\infty}^{\infty} \cdots \sum_{k_d=-\infty}^{\infty}$$

$$X(k_1, k_2, \ldots, k_d) e^{\mathbf{j}2\pi s\left[\frac{k_1\lambda_1}{\lambda_d} + \cdots + \frac{k_{d-1}\lambda_{d-1}}{\lambda_d}\right]} e^{\mathbf{j}k_d\lambda_d\theta} e^{\mathbf{j}\psi}$$

Let $Y(k_1, k_2, \ldots, k_{d-1}) = \sum_{k_d=-\infty}^{\infty} X(k_1, k_2, \ldots, k_d) e^{\mathbf{j}k_d\lambda_d\theta}$

$$x_s = \sum_{k_1=-\infty}^{\infty} \cdots \sum_{k_{d-1}=-\infty}^{\infty} Y(k_1, k_2, \ldots, k_{d-1}) e^{\mathbf{j}2\pi s\left[\frac{k_1\lambda_1}{\lambda_d} + \cdots + \frac{k_{d-1}\lambda_{d-1}}{\lambda_d}\right]} e^{\mathbf{j}\psi}$$

$\square$

The MFT method uses the quasiperiodic sampling theorem to insure that the waveforms are d-quasiperiodic by sampling them at the clock frequency and insisting that the sampled waveforms be $(d-1)$-quasiperiodic. This is an easier condition to work with because the quasiperiodic constraint is placed on a discrete waveform with a rather long time between samples as opposed to a continuous waveform. To make this task tractable, it is necessary to assume that the Fourier series of the sampled waveform has only a finite number of nonzero terms. It is always possible to approach this ideal arbitrarily closely because any signal present in a physical system must have finite bandwidth. However, for the MFT method to be practical, there must be only a small number of nonzero terms in the Fourier series.

Based on Theorem 7.1, the sampled signals can be represented by a Fourier series with $d-1$ fundamentals. Assume that only J terms in this Fourier series are nonzero, then knowing the value of J samples allows the Fourier series of the sampled waveforms to be computed. Once the coefficients of the Fourier series are known, the value of any sample could be computed. In particular, given the value of J samples, it is possible to compute what must be the value of the immediately following J samples for the assumption of a finite Fourier series to hold. In other words, the finite Fourier series assumption yields a relationship between any J non-adjacent points on the sampled waveforms and their immediate successors, which is referred to as the delay operator. Another such relationship exists through the differential equations that describe the circuit. For each of the J samples, the equations are integrated over an interval of one clock cycle, each time using a different one of the J samples as an initial state. Thus, using the differential equations, it is also possible to start with J points on the sampled waveform and compute the immediately following points, as illustrated in Figure 7.2. MFT uses these two relationships to compute the quasiperiodic response of a circuit by finding the value of the J prespecified non-adjacent samples for which these two relationships agree. If the two relationships agree and if the assumption of only J nonzero terms in the Fourier series holds, then the entire waveform is a d-quasiperiodic solution to the differential equations.

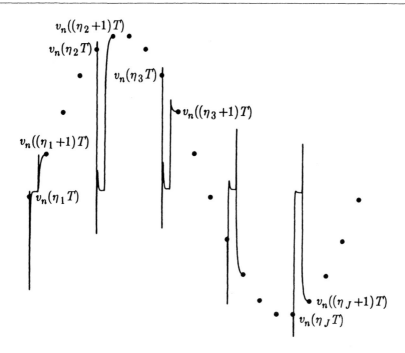

Figure 7.2: The discrete waveform that is constructed by sampling the response of a circuit at the initial point of each clock cycle. Also shown are the J cycles that are calculated in detail.

2.1. The Delay Operator

Consider the sequence of points generated by sampling the quasi-periodic response of the lumped test problem (3.11), and denote the sequence $\{v(sT) : s \in \mathbf{Z}\}$ where $T = 2\pi/\lambda_d$ and λ_d is considered the clock. From the quasiperiodic sampling theorem the Fourier series for this sequence is given by

$$v(sT) = V_0 + \sum_{k_1=1}^{\infty} \cdots \sum_{k_{d-1}=1}^{\infty}$$

$$\left[V^C(k_1, \ldots, k_{d-1})\cos(sT(k_1\lambda_1 + \cdots + k_{d-1}\lambda_{d-1})) + \right.$$

$$\left. V^S(k_1, \ldots, k_{d-1})\sin(sT(k_1\lambda_1 + \cdots + k_{d-1}\lambda_{d-1})) \right]$$

where the clock phase θ has been dropped to simplify the notation. Assume the sequence can be accurately approximated by the truncated Fourier series that results from considering only the first H harmonics of each fundamental (this is the box truncation of Chapter 3). The set of frequencies is given by

$$\Lambda_K = \{\omega : \omega = k_1\lambda_1 + \cdots + k_d\lambda_d;$$

$$|k_j| = 0, 1, \ldots, H \text{ for } 1 \leq j \leq d; \text{ first nonzero } k_j \text{ positive}\}.$$

There are K nonzero frequencies in this series and $J = 2K + 1$ unknown coefficients. Assuming all frequencies are distinct, there is a linear relation between any collection of J initial points and any other collection of J initial points. However, as mentioned above, we are most interested in the linear operator that maps a collection $v(\eta_1 T), \ldots, v(\eta_J T)$ into $v((\eta_1+1)T), \ldots, v((\eta_J+1)T)$ where T is the clock period and $\{\eta_1, \ldots, \eta_J\} \subset Z$. This linear operator is referred to as the delay matrix.

    Deriving the delay matrix is a two stage process. First, the J points, $v(\eta_1 T), \ldots, v(\eta_J T)$ are used to calculate the Fourier coefficients. Then the Fourier series (using these coefficients) is evaluated at the J times, $(\eta_1+1)T, \ldots, (\eta_J+1)T$. The Fourier coefficients are then eliminated to yield the desired direct relation. To compute the Fourier coefficients, write the truncated Fourier series as a system of J linear equations in J unknowns using the APFT,

$$
\Gamma_0^{-1}
\begin{bmatrix}
V_n(0) \\
V_n^C(1) \\
V_n^S(1) \\
\vdots \\
V_n^C(K) \\
V_n^S(K)
\end{bmatrix}
=
\begin{bmatrix}
v_n(\eta_1 T) \\
v_n(\eta_2 T) \\
v_n(\eta_3 T) \\
\vdots \\
v_n(\eta_J T)
\end{bmatrix}.
\tag{7.4}
$$

where $\Gamma_0^{-1} \in \mathbb{R}^{J \times J}$ is given by

$$
\Gamma_0^{-1} =
\begin{bmatrix}
1 & \cos\omega_1\eta_1 T & \sin\omega_1\eta_1 T & \cdots & \cos\omega_K\eta_1 T & \sin\omega_K\eta_1 T \\
1 & \cos\omega_1\eta_2 T & \sin\omega_1\eta_2 T & \cdots & \cos\omega_K\eta_2 T & \sin\omega_K\eta_2 T \\
1 & \cos\omega_1\eta_3 T & \sin\omega_1\eta_3 T & \cdots & \cos\omega_K\eta_3 T & \sin\omega_K\eta_3 T \\
\vdots & \vdots & \vdots & & \vdots & \vdots \\
1 & \cos\omega_1\eta_J T & \sin\omega_1\eta_J T & \cdots & \cos\omega_K\eta_J T & \sin\omega_K\eta_J T
\end{bmatrix}
$$

where $\omega_k \in \Lambda_K$. The matrix Γ_0^{-1} maps the Fourier coefficients to a sequence and is referred to as the inverse almost-periodic Fourier transform. Its inverse, if it exists, is the forward almost-periodic Fourier transform and is denoted by Γ_0. We can also write

$$
\Gamma_T^{-1}
\begin{bmatrix}
V_n(0) \\
V_n^C(1) \\
V_n^S(1) \\
\vdots \\
V_n^C(K) \\
V_n^S(K)
\end{bmatrix}
=
\begin{bmatrix}
v_n((\eta_1+1)T) \\
v_n((\eta_2+1)T) \\
v_n((\eta_3+1)T) \\
\vdots \\
v_n((\eta_J+1)T)
\end{bmatrix}
\tag{7.5}
$$

where $\Gamma_T^{-1} \in \mathbb{R}^{J \times J}$ is given by

$$
\Gamma_T^{-1} = \begin{bmatrix}
1 & \cos\omega_1(\eta_1+1)T & \sin\omega_1(\eta_1+1)T & \cdots & \cos\omega_K(\eta_1+1)T \\
1 & \cos\omega_1(\eta_2+1)T & \sin\omega_1(\eta_2+1)T & \cdots & \cos\omega_K(\eta_2+1)T \\
1 & \cos\omega_1(\eta_3+1)T & \sin\omega_1(\eta_3+1)T & \cdots & \cos\omega_K(\eta_3+1)T \\
\vdots & \vdots & \vdots & & \vdots \\
1 & \cos\omega_1(\eta_J+1)T & \sin\omega_1(\eta_J+1)T & \cdots & \cos\omega_K(\eta_J+1)T
\end{bmatrix}
$$

$$
\begin{bmatrix}
\sin\omega_K(\eta_1+1)T \\
\sin\omega_K(\eta_2+1)T \\
\sin\omega_K(\eta_3+1)T \\
\vdots \\
\sin\omega_K(\eta_J+1)T
\end{bmatrix}.
$$

Given a sequence, a delayed version is computed by applying Γ_0 to the sequence to compute the Fourier coefficients, and then multipling the vector of coefficients by Γ_T^{-1}.

$$
\begin{bmatrix}
v_n((\eta_1+1)T) \\
v_n((\eta_2+1)T) \\
\vdots \\
v_n((\eta_J+1)T)
\end{bmatrix}
= \Gamma_T^{-1}\Gamma_0
\begin{bmatrix}
v_n(\eta_1 T) \\
v_n(\eta_2 T) \\
\vdots \\
v_n(\eta_J T)
\end{bmatrix}
\tag{7.6}
$$

Thus, the delay matrix, $D(T) \in \mathbb{R}^{J \times J}$, is defined as

$$
D(T) = \Gamma_T^{-1}\Gamma_0. \tag{7.7}
$$

As the delay matrix is a function only of $\{\omega_1, \omega_2, \ldots, \omega_K\}$, $\{\eta_1, \eta_2, \ldots, \eta_J\}$, and T, it is computed once and used for every node.

2.2. The Differential Equation Relation

If the node voltages are known at some time t_0, it is possible to solve the equation describing the lumped test problem (3.11) and compute the node voltages at some later time t_1. The relationship between $v(t_0)$

and $v(t_1)$ is expressed using the state-transition function

$$v(t_1) = \phi(v(t_0), t_0, t_1) \tag{7.8}$$

where ϕ can be expanded as

$$\phi(v(t_0), t_0, t_1) = \begin{bmatrix} \phi_1(v(t_0), t_0, t_1) \\ \phi_2(v(t_0), t_0, t_1) \\ \vdots \\ \phi_N(v(t_0), t_0, t_1) \end{bmatrix} \tag{7.9}$$

where $\phi_n : \mathbb{R}^{N \times 1 \times 1} \to \mathbb{R}$ for all circuit nodes $n = 1, 2, \ldots, N$.

Now reconsider the J initial points at some circuit node n, $v_n(\eta_1 T), \ldots, v_n(\eta_J T)$. For each $j = 1, 2, \ldots, J$ and each $n = 1, 2, \ldots, N$ write

$$v_n((\eta_j + 1)T) = \phi_n(v(\eta_j T), \eta_j T, (\eta_j + 1)T) \tag{7.10}$$

where T is the clock period. Note that $v_n((\eta_j + 1)T)$ is the initial point of the cycle immediately following the cycle beginning at $\eta_j T$. Also, the node voltages at $\eta_j T$ are related to the node voltages at $(\eta_j + 1)T$ by the delay matrix, $D(T)$. That is,

$$D(T) \begin{bmatrix} v_n(\eta_1 T) \\ v_n(\eta_2 T) \\ \vdots \\ v_n(\eta_J T) \end{bmatrix} = \begin{bmatrix} v_n((\eta_1 + 1)T) \\ v_n((\eta_2 + 1)T) \\ \vdots \\ v_n((\eta_J + 1)T) \end{bmatrix} \tag{7.11}$$

It is possible to use (7.10) to eliminate the $v_n((\eta_j + 1)T)$ terms from (7.11), which yields

$$D(T) \begin{bmatrix} v_n(\eta_1 T) \\ v_n(\eta_2 T) \\ \vdots \\ v_n(\eta_J T) \end{bmatrix} = \begin{bmatrix} \phi_1(v(\eta_1 T), \eta_1 T, (\eta_1 + 1)T) \\ \phi_2(v(\eta_2 T), \eta_2 T, (\eta_2 + 1)T) \\ \vdots \\ \phi_N(v(\eta_J T), \eta_J T, (\eta_J + 1)T) \end{bmatrix} \tag{7.12}$$

for each node $n = 1, 2, \ldots, N$.

Solving the N simultaneous equations (7.12) results in knowing $v(\eta_j T)$ for $j = \eta_1, \eta_2, \ldots, \eta_J$. From these numbers, the Fourier coefficients $V(0), V(1), \ldots, V(K)$ can be computed. Once the Fourier coefficients are known, it is possible to compute $v(jT)$ for any integer j. The value of $v(t)$ for any t is found by integrating (3.11) using the nearest preceding $v(jT)$ as a starting point.

2.3. An Example

Consider the simple switched-capacitor RC one-pole filter shown in Figure 7.3. It is easy to show that

$$\phi_1(v(\eta_j T), \eta_j T, (\eta_j + 1)T) = \frac{C_1 v_{in}(\eta_j T) + C_2 v(\eta_j T)}{C_1 + C_2}.$$

Since the circuit is linear, it is only necessary to consider DC and the fundamental in the solution, and so only three samples are needed. Assume the fundamental is 1 Hz and the clock is 6 Hz, and choose the samples to be taken at $t = \{0, 1/3, 2/3\}$. Then

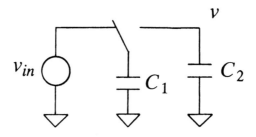

Figure 7.3: A simple switched-capacitor RC filter.

$$\Gamma_0^{-1} = \begin{bmatrix} 1 & \cos(0) & \sin(0) \\ 1 & \cos(\frac{2\pi}{3}) & \sin(\frac{2\pi}{3}) \\ 1 & \cos(\frac{4\pi}{3}) & \sin(\frac{4\pi}{3}) \end{bmatrix} = \begin{bmatrix} 1 & 1 & 0 \\ 1 & -\frac{1}{2} & \frac{\sqrt{3}}{2} \\ 1 & -\frac{1}{2} & -\frac{\sqrt{3}}{2} \end{bmatrix},$$

$$\Gamma_0 = \begin{bmatrix} \frac{1}{3} & \frac{1}{3} & \frac{1}{3} \\ \frac{2}{3} & -\frac{1}{3} & -\frac{1}{3} \\ 0 & \frac{1}{\sqrt{3}} & -\frac{1}{\sqrt{3}} \end{bmatrix},$$

$$\Gamma_T^{-1} = \begin{bmatrix} 1 & \cos(\frac{\pi}{3}) & \sin(\frac{\pi}{3}) \\ 1 & \cos(\pi) & \sin(\pi) \\ 1 & \cos(\frac{5\pi}{3}) & \sin(\frac{5\pi}{3}) \end{bmatrix} = \begin{bmatrix} 1 & \frac{1}{2} & \frac{\sqrt{3}}{2} \\ 1 & -1 & 0 \\ 1 & \frac{1}{2} & -\frac{\sqrt{3}}{2} \end{bmatrix},$$

and

$$D(T) = \Gamma_T^{-1}\Gamma_0 = \frac{1}{3} \begin{bmatrix} 2 & 2 & -1 \\ -1 & 2 & 2 \\ 2 & -1 & 2 \end{bmatrix}.$$

Substituting into (7.12),

$$\frac{1}{3} \begin{bmatrix} 2 & 2 & -1 \\ -1 & 2 & 2 \\ 2 & -1 & 2 \end{bmatrix} \begin{bmatrix} v(0) \\ v(\frac{1}{3}) \\ v(\frac{2}{3}) \end{bmatrix} = \frac{1}{C_1 + C_2} \begin{bmatrix} C_1 v_{in}(0) + C_2 v(0) \\ C_1 v_{in}(\frac{1}{3}) + C_2 v(\frac{1}{3}) \\ C_1 v_{in}(\frac{2}{3}) + C_2 v(\frac{2}{3}) \end{bmatrix}.$$

By giving values for C_1, C_2, and $v_{in}(sT)$ for $s = 0, 2, 4$, this linear equation is solved for $[v(0) \ v(\frac{1}{3}) \ v(\frac{2}{3})]^T$. The Fourier coefficients are computed with

$$\left[V(0) \quad V^C(1) \quad V^S(1) \right] = \Gamma_0 \begin{bmatrix} v(0) \\ v(\frac{1}{3}) \\ v(\frac{2}{3}) \end{bmatrix}.$$

2.4. MFT as a Two-Point Boundary Constraint

To show that the MFT formulation is interpretable as a two-point boundary-value problem, the time for each of the J intervals is written as a function of a new independent variable τ that ranges over the interval $[0, 1]$. The boundaries for the two-point boundary-value problem are taken to be $\tau = 0$ and $\tau = 1$. For interval j, $t_j = (\eta_j + \tau)T$ and let

$$V(\tau) = [v((\eta_1 + \tau)T), v((\eta_2 + \tau)T), \dots, v((\eta_J + \tau)T)]^{\mathrm{T}}$$

and so $V(\tau) \in \mathbb{R}^{NJ}$. Consider the set of differential equations that describe the circuit over one cycle to be independent from the set that describe the circuit in any other cycle and combine the J independent sets into one composite set of differential equations for which V is the state vector. Denote Φ as the composite state-transition function for the enlarged set of equations. The differential equation relation becomes

$$V(1) = \Phi(V(0), 0, 1) = \begin{bmatrix} \phi(v(\eta_1 T), \eta_1 T, (\eta_1 + 1)T) \\ \phi(v(\eta_2 T), \eta_2 T, (\eta_2 + 1)T) \\ \vdots \\ \phi(v(\eta_J T), \eta_J T, (\eta_J + 1)T) \end{bmatrix}. \qquad (7.13)$$

The delay equation is written

$$V(1) = D_N(T)V(0). \qquad (7.14)$$

$D_N(T) \in \mathbb{R}^{NJ \times NJ}$ is the composite delay operator and is given by

$$D_N(T) = \begin{bmatrix} d_{11}\mathbf{1}_N & \cdots & d_{1J}\mathbf{1}_N \\ d_{J1}\mathbf{1}_N & \cdots & d_{JJ}\mathbf{1}_N \end{bmatrix} \qquad (7.15)$$

where $d_{ij} \in \mathbb{R}$ is the ijth element of the delay matrix $D(T)$ and $\mathbf{1}_N \in \mathbb{R}^N$ is the identity matrix. Equation (7.14) is a two-point boundary constraint on (7.13). The solution of these two simultaneous equations is a collection of values that fall on a quasiperiodic solution for the lumped test problem at the prespecified points. These two equations can be combined into one by eliminating $V(1)$.

$$\Phi(V(0), 0, 1) - D_N(T)V(0) = 0 \qquad (7.16)$$

This equation is equivalent to (3.20) with $A = -D_N(T)$, $B = \mathbf{1}_{NJ}$, $c = 0$, $\phi = \Phi$, and $x = V$.

3. Practical Issues

MFT poses the problem of finding a quasiperiodic solution to the lumped test problem (3.11) as a boundary-value problem by using (7.13) and (7.14). These equations can be solved by using either finite-difference methods or shooting methods (Chapter 4). In either case, the number of unknowns, and hence the time required to compute the solution, is likely to be large. With shooting methods and for a circuit with N nodes, the number of unknowns is NJ, where J is the number of samples required by MFT. Recall that MFT uses the quasiperiodic sampling theorem to convert the problem of finding a continuous time d-quasiperiodic solution to the problem of finding a discrete-time $(d-1)$-quasiperiodic solution. J is the number of unknown terms in the Fourier series of the discrete-time $(d-1)$-quasiperiodic signal. To reduce the time required to compute the solution, both the structure of the MFT must be exploited during the computations required by the shooting method and J, the number of terms in the Fourier series, must be minimized.

3.1. The Quasiperiodic Sampled Waveform

The number of unknowns in MFT is proportional to the number of significant terms in the Fourier series of the $(d-1)$-quasiperiodic sampled waveform. There are two degrees of freedom that can be exploited to reduce the number of significant terms in this series. First, any one of the d fundamental frequencies can be chosen as the clock. This is discussed further in the next section. Second, the phase of the clock signal at which the samples are taken (θ in the quasiperiodic sampling theorem) is as yet unspecified. It can be advantageous to choose this phase judiciously when the clock signal is causing the circuit to switch. For example, consider the circuit shown in Figure 7.4. The voltage at the output and summing nodes is sampled with $\theta = 0$ and $\theta = 0.25T$. The spectra of the sampled waveforms are shown in Figure 7.5. This shows that the number of significant terms in the Fourier series of the sampled signal is much higher when $\theta = 0.25T$. This behavior results because at $\theta = 0$ there is a delay of T seconds between when the switch last changed state and when the waveforms are sampled. Conventional design practice dictates that any transients should have decayed to negligible levels during this interval. At this time, feedback in the circuit is acting to minimize distortion. Conversely, when $\theta = 0.25T$ the samples are being taken at a time when the op amp is undergoing slew rate limiting, an effect that results from nonlinearities in the input stage.

It is very common for analog switching circuits to respond very nonlinearly to the clock signal but respond (by design) nearly linearly to other input signals. Choosing the proper clock phase to sample the waveforms judiciously serves to reduce computation time in two important ways. First, it can greatly reduce the number of Fourier series terms needed to represent the sampled signal accurately. Second, choosing θ to reduce J, also serves to make Φ in (7.13) more linear and so reduces the number of iterations required by the shooting method (Chapter 4).

3.2. Selecting the Clock

In the MFT method the circuit equations are integrated for a total of J clock cycles. The number of time-points required in one clock period is proportional to $T\lambda_d$, where λ_d is the largest of the fundamental frequencies and T is the clock period. Thus, the total number of time-

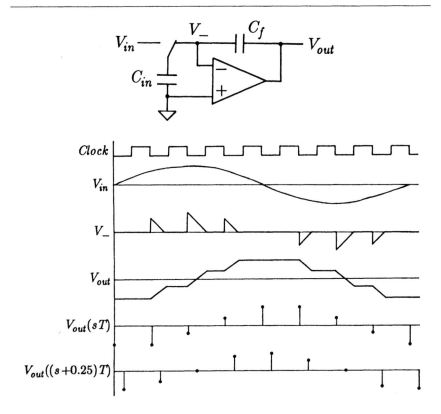

Figure 7.4: A switched capacitor integrator and its steady-state response to a sinusoidal input.

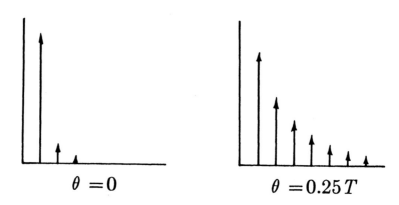

Figure 7.5: Spectra for the sampled waveforms of Figure 7.4.

points required is proportional to $JT\lambda_d$ where λ_d is fixed. This forms the basis used to choose which fundamental frequency should be the clock: the clock is chosen to be the fundamental that minimizes the JT product. Clearly, the T term is minimized by selecting the largest fundamental to be the clock. However, if the largest fundamentals are close in frequency, it is sometimes desirable to use other fundamentals as the clock because a smaller J can be used. For example, in a circuit that exhibits a d-quasiperiodic solution with λ_{d-1} close to, but less than, λ_d, it would be preferable to choose λ_{d-1} as the clock if either the input sources had considerably larger signal levels at frequencies λ_{d-1} and its harmonics than at other frequencies, or if the state-transition function of the circuit for one period of the fundamental λ_{d-1} is considerably more linear than for one period of the other fundamentals. If either or both situations are present, then J could be smaller if λ_{d-1} is chosen as the clock over λ_d.

4. Nitswit

Nitswit is a circuit simulator developed duing the course of this research that implements the MFT method and solves for quasiperiodic steady-state solutions [kundert88c] [kundert89]. The name results from *Nitswit* being a detailed (nit) level circuit simulator for analog switching (swit) circuits. *Nitswit* is the first circuit simulator capable of finding quasiperiodic steady-state solutions directly in the time domain. It uses the MFT method to formulate the problem of finding a quasi-periodic steady-state solution as a boundary-value problem and Newton-Raphson-based shooting methods to solve the boundary-value problem. *Nitswit* exploits the property of shooting methods that allows them to easily a handle circuit with a near-linear state-transition function over the shooting interval even when the circuit is behaving very nonlinearly during the interval. This allows *Nitswit* to handle analog switching circuits such as switching mixers and switched-capacitor filters. Even the difficult task of computing the intermodulation distortion of narrow-band switched-capacitor filters is performed efficiently with *Nitswit*. It is designed to simulate these circuits without the approximations of the discrete-time methods of Fang [fang83] and Rabaey [de man80] (such as the slow-clock and macromodeling approximations).

4.1. Equation Formulation

Nitswit applies shooting methods to solve the mixed frequency-time formulation equations (7.13) and (7.14), rewritten slightly here as

$$F \begin{bmatrix} v(\eta_1 T) \\ v(\eta_2 T) \\ \vdots \\ v(\eta_J T) \end{bmatrix} = D_N(T) \begin{bmatrix} v(\eta_1 T) \\ v(\eta_2 T) \\ \vdots \\ v(\eta_J T) \end{bmatrix} - \begin{bmatrix} \phi(v(\eta_1 T), \eta_1 T, (\eta_1 + 1)T) \\ \phi(v(\eta_2 T), \eta_2 T, (\eta_2 + 1)T) \\ \vdots \\ \phi(v(\eta_J T), \eta_J T, (\eta_J + 1)T) \end{bmatrix} = 0$$

(7.19)

where $F : \mathbb{R}^{NJ} \to \mathbb{R}^{NJ}$, and $D_N(T)$ is given by

$$D_N(T) = \begin{bmatrix} d_{11}\mathbf{1}_N & \cdots & d_{1J}\mathbf{1}_N \\ d_{J1}\mathbf{1}_N & \cdots & d_{JJ}\mathbf{1}_N \end{bmatrix}$$

(7.20)

where $d_{ij} \in \mathbb{R}$ is the ijth element of the delay matrix $D(T)$ and $\mathbf{1}_N \in \mathbb{R}^N$ is the identity matrix. This equation is solved with Newton-Raphson based shooting methods as shown in Chapter 4.

4.2. Implementation

Both the classical direct methods and the mixed frequency-time methods have been implemented in the simulation program *Nitswit*, which is written in the computer language "C." *Nitswit* takes as input a file with a SPICE-like description of the circuit, that is, a list of elements (MOS transistors, resistors, capacitors, etc) with their node connections, and a list of options to select among methods. If the mixed frequency-time method is used, a switching clock period and an input frequency, along with a number of harmonics, must be specified. The program produces as output waveforms as in Figure 7.1 for direct methods, and waveforms as in Figure 7.2 and Fourier coefficients for the sampled waveforms with the mixed frequency-time algorithm.

4.2.1. Application Examples

Nitswit is particularly efficient when simulating switched-capacitor filters. One reason is that switched-capacitor filters are usually followed by a sampler, and so only the initial point of each cycle is needed. Also, the circuits are generally designed to exhibit little distortion, so if driven by a sinusoid, only a few harmonics of the sequence of initial points are significant and only a few full clock cycles need to be computed. Finally, the state-transition function for a switched-capacitor filter over a clock cycle is nearly affine (linear plus a constant), and therefore Newton-Raphson applied to (7.19) converges in just a few iterations.

To demonstrate the effectiveness and versatility of the algorithms used in *Nitswit*, we consider analyzing the distortion of a switched-capacitor circuit. Figure 7.6 shows a high-speed fully-differential switched-capacitor sample-and-hold amplifier [lewis87]. This circuit precedes an analog-to-digital converter and has all three characteristics mentioned above. An important performance specification for this circuit is distortion. Its distortion is measured by applying a sinewave to the input and a periodic clock to the sample/hold input. The output signal is then sampled at the end of each hold interval and a Fourier series is constructed from the sampled signal. If the sample-and-hold is ideal, there will be energy only at the frequency of the input sinusoid. Any energy at other frequencies is considered distortion. The sampling of the output at the end of the hold interval is needed to eliminate settling effects that result at transitions of the sample/hold signal that are ignored by the analog-to-digital converter that follows. Conventional circuit simulators limit the input frequency to be near the sample/hold clock rate and do not sample the output signal at the end of the hold interval before computing its Fourier series.

Figure 7.7 shows the operational amplifier used in the sample-and-hold of Figure 7.6. The combined circuit contains 65 nodes. The distortion of this circuit was measured with *Nitswit* versus the amplitude of the input signal (Figure 7.8) and versus the sample/hold clock frequency (Figure 7.9). Of particular interest to the designer is the distortion versus clock rate. This is a quantity that cannot be determined except with a circuit-level simulator such as *Nitswit* or by measuring the actual circuit.

These simulation were performed with a SPICE level 1 MOS model with a simplified version of the charge conserving model of Yang and Chatterjee [yang82]. This model was chosen simply because it was easy to implement. However, it is generally considered to be

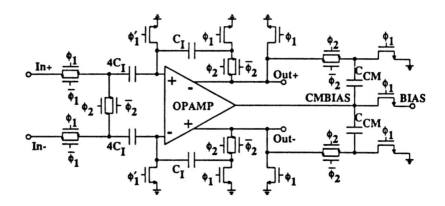

Figure 7.6: A full-differential switched-capacitor sample-and-hold amplifier (courtesy of Steve Lewis [lewis87]).

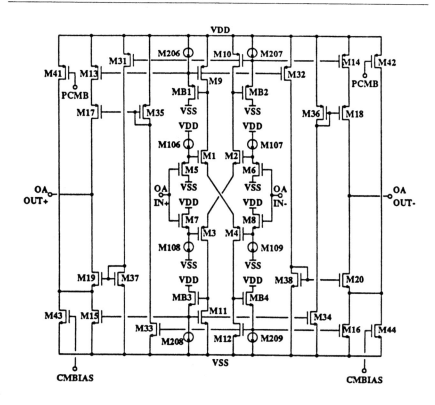

Figure 7.7: The operational amplifier used in the sample-and-hold amplifier of Figure 7.6 (courtesy of Steve Lewis [lewis87]).

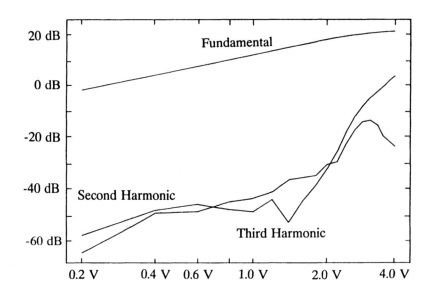

Figure 7.8: Distortion of the sample-and-hold amplifier as a function of input signal amplitude. The input signal frequency is 500 kHz and the sample/hold clock rate was 10 MHz.

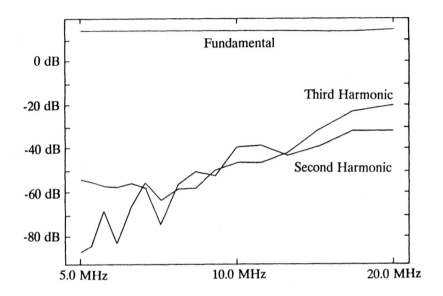

Figure 7.9: Distortion of the sample-and-hold amplifier as a function of sample/hold clock frequency. The amplitude of the input signal is 1.25 V differential and its frequency is 500 kHz.

too inaccurate to be suitable for analog circuits. Since the mixed frequency-time algorithm has extra overhead (primarily, the time required to construct the sensitivity matrix) that dominates over the time required for model evaluation, it is expected that the algorithm should do better with respect to direct methods when a more accurate (and therefore a more complicated) model is used.

The Fourier series for the sampled signal was truncated after three harmonics. For the case where the input was a 2 V differential 500 kHz sine wave and the sample/hold clock rate was 10 MHz, this gave results that were identical to direct methods to within the truncation error of the integration method.

4.2.2. Comparison to Direct Methods

The program *Nitswit* contains two algorithms capable of finding the steady-state response of a circuit. The first is simply a transient analysis that continues until a steady-state is achieved. The second, of course, is the mixed frequency-time algorithm. Coding both algorithms into the same simulator provides a fair evaluation of the mixed frequency-time approach.

Figure 7.10 shows the time required to simulate the sample-and-hold of Figure 7.6 to steady-state as a function of the input signal frequency for both direct methods and for the mixed-frequency algorithm. The amplitude of the input signal is 1.25 V differential and its frequency is 500 kHz. The sample/hold clock is fixed at 5 MHz and three harmonics of the sampled signal are computed. This figure shows that the time required for the mixed frequency-time algorithm is roughly independent of the frequency of the input signal whereas the time required for direct method is proportional to the ratio of the clock frequency to the input frequency. This circuit provides the freedom of choosing the input frequency close to the clock frequency, allowing transient analysis to be efficient. Many circuits do not provide this freedom.

Results for four circuits are given in Table 7.1. The first, *sclpf*, is an RC one-pole SC filter. The second, *scop*, is a one-pole active CMOS low pass filter. The circuit, *mixer*, is a double-balanced switching mixer with a 1.001 MHz RF input signal and a 1 MHz LO signal. This circuit shows that *Nitswit* is not limited to switched-capacitor circuits. The last circuit, *frog*, is a five pole Chebyshev active CMOS leap frog filter with 0.1 dB ripple. This circuit is driven with a 1 MHz clock, has a 20 kHz bandwidth, and is being driven with a 1 kHz test signal to measure its distortion.

Examination of the results above indicate as much as an order of magnitude speed increase over traditional methods, but this is not as much as one would expect. Much of the CPU time for large circuits, such as *frog*, is spent calculating the dense sensitivity matrix and

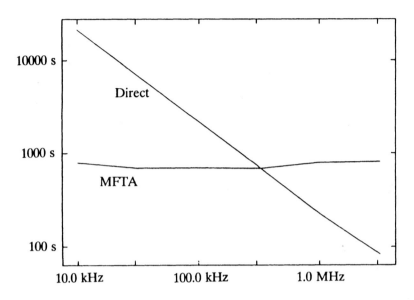

Figure 7.10: Time required for simulation of sample-and-hold amplifier to steady state versus input frequency using direct methods and the mixed frequency-time algorithm. These time were measured on an HP9000/370 with a floating point accelerator.

Circuit			Direct	Mixed Frequency-Time			Ratio
Name	Nodes	Cycles/ Period	Time (sec)	Harms	Newton Iters	Time (sec)	Direct/ MFT
sclpf	2	33	24.5	3	3	4.3	5.7
scop	13	100	522	3	6	90	5.8
mixer	34	1000	7132	3	4	161	44.3
frog	77	1000	12,987	3	6	1228	10.6

Table 7.1: *Nitswit* results from a VAX 8650 running ULTRIX 2.0.

factoring the Jacobian. It does turn out however, that almost all the entries of the sensitivity matrix are near zero, and this suggests significant speed improvements can be achieved by ignoring those terms.

Chapter 8
Comparisons

For each of the three classes of steady-state methods presented, finite-difference, shooting, and harmonic balance methods, there are situations where each is best. In this chapter, each method will be summarized and its advantages and disadvantages given. Each of the circuits presented in Chapter 2 are also reviewed and recommendations are given on how to find their steady-state response most efficiently.

1. Finite-Difference Methods

Finite-difference methods solve boundary-value problems by discretizing the differential equation for the circuit on a finite set of time-points that cover the simulation interval. This results in a large system of equations which are solved for the node voltages at each time-point simultaneously. The important characteristics of this method are that the equation and solution are formulated in the time domain, it solves boundary-value problems, and it solves for the whole solution simultaneously.

The nonlinear finite-difference equations are usually solved using Newton-Raphson. Since there are large number of equations and unknowns, the Jacobian is quite large. This is a significant drawback. This method cannot be applied to large circuits because of the large amount of memory that is required for the Jacobian. The matrix is sparse, which helps considerably.

When Newton-Raphson is applied to equations generated by the finite-difference method, the intermediate iterates are waveforms that may not satisfy the differential equations, but do satisfy the boundary constraints (assuming that these constraints are linear). For this reason, the finite-difference methods are well suited for finding quasiperiodic or unstable periodic solutions. This last aspect may or may not be a feature of the method, depending on whether you wish to find or avoid unstable solutions. The finite-difference methods are like parallel shooting methods in this regard. In fact, a parallel shooting method, when the shooting intervals are taken to be so small that there is one time-step per interval, becomes a finite-difference method. This is easily seen by using the one step approximation $\phi_k(t_{k-1}) = x_k + \dfrac{1}{h_k}\dot{x}_k$ in equations (4.31), which becomes a finite-difference method with explicit Euler used to form the difference equations.

Finite-difference methods allow the most freedom when choosing an integration method. The time-steps may be nonuniformly spaced and the spacing may be chosen to increase accuracy by clustering points where waveforms are changing rapidly and increasing efficiency by spreading them out where the waves are quiescent. Furthermore, finite-difference methods allow the integration method to be noncausal. In other words, the value of the solution at future time-points as well as past time-points are used when computing the discretized derivative. Noncausal discretization can be used with finite-difference methods because the solution is calculated for all time at once. No attempt is made to exploit temporal unidirectionality.

Finite-difference methods are able to solve boundary-value problems formulated using the MFT formulation, and so are able to find quasiperiodic solutions with an arbitrary discretization method. They are also able to handle circuits with distributed devices, and like all steady-state methods, exhibit characteristics that can be exploited on parallel or vector machines.

2. Shooting Methods

Shooting methods convert boundary-value problems into a sequence of initial-value problems. They begin with a guess at the solution at the beginning of the shooting interval. The equations are integrated over the interval and the results at each of the boundaries are substituted into the boundary constraint and the initial state is corrected in a manner to better satisfy the boundary constraint. The guess is corrected using either an extrapolation method or a Newton-Raphson method. Unlike finite-difference methods, the intermediate iterates generated by shooting methods satisfy the differential equations, but not the boundary constraint.

Shooting methods exploit the causality inherent in most dynamic systems by evaluating the differential equation as an initial-value problem. They do not compute the solution to the differential equation all at once and only need access to a small piece of it (a few past time-points) at any point in time. As a result, they need considerably less memory.

Shooting methods have limited utility on large circuits however. With Newton-Raphson based shooting methods, a matrix equation involving the Jacobian must be solved at each time-point (solving this equation requires N forward/backward substitutions of the Jacobian, where N is the size of the Jacobian). The computational complexity of this step is at best $O(N^2)$ (for a diagonal Jacobian) and at worst $O(N^3)$ (for a full Jacobian). The matrix equation is solved when updating the sensitivity matrix. The sensitivity matrix generally exhibits considerably numerical sparsity[1], however the sparsity pattern is not known in advance and changes during the course of the computation. How to exploit this sparsity is an open question that, if solved, would result in Newton-Raphson based shooting becoming considerably more efficient on large circuits.

---

[1]The sensitivity matrix is numerically sparse in the sense that there are many entries in the matrix that are close to zero. This is not the usual definition of sparsity, where many entries are exactly zero.

Extrapolation based shooting methods need to evaluate the differential equation over p shooting intervals, where p is the number of time constants on the order of, or greater than, the shooting interval. For large circuits, p could be quite large, but that would be unusual. However, simulating a large circuit over many shooting intervals for each iteration of the shooting method can get quite expensive.

A very important characteristic of shooting methods is that they converge quickly and reliably if the state-transition function is near linear. It is quite often the case (usually by design) that the state-transition function is linear even when the overall circuit behavior is not. The nonlinear behavior is not a problem for the numerical integration used to evaluate ϕ_T because numerical integration is a natural continuation method where time is the continuation parameter. This hiding of the nonlinear behavior gives shooting methods a considerable advantage over the finite-difference methods and harmonic balance on a wide range of problems.

An important disadvantage for shooting methods is that they cannot handle distributed devices. Circuits containing distributed devices could be solved with shooting methods if the distributed devices where replaced with lumped approximations, however the lumped approximation often contains so many internal nodes and significant time constants, that they considerably increase the cost of using shooting methods.

Lastly, shooting methods can find the quasiperiodic solution of a circuit if the boundary-value problem is formulated using the MFT method. Shooting methods can be accelerated using parallel and vector machines, and particularly so if parallel shooting methods are used. However, the advantage that shooting methods enjoy by hiding nonlinear behavior from the outer loop is often lost with parallel shooting methods.

3. Harmonic Balance

Harmonic balance is a frequency-domain method that is unique in two ways. First, the equations are formulated by assuming that the voltages and currents in the circuit are approximated by a Fourier series accurately. Since such signals are quasiperiodic, the method inherently avoids transient behavior by being incapable of representing it. The complexity of the method is determined by the number of terms needed in the Fourier series to accurately represent the signals, and not by the actual frequencies. The complexity is low and the method efficient if the circuit is behaving near linearly and is driven by signals representable by Fourier series with few terms. In this situation, a truncated Fourier series is often a very good approximation to the true solution, and harmonic balance is very accurate. Indeed, an important feature of harmonic balance is that for near-linear circuits being driven with a sinusoid, harmonic balance becomes more accurate as the input signal level is reduced and the circuit becomes more linear. In the limit, harmonic balance is exact for linear circuits.

The second way that harmonic balance is different from time-domain methods is that it represents signals using the coefficients of the sinusoids rather than a sampled-data representation (i.e., in the frequency domain using the Fourier series rather than in the time domain). Using Fourier coefficients and superposition allows the linear components to be evaluated with phasor analysis. Thus, the large number of distributed device models available in the literature that are formulated using phasors are directly compatible with harmonic balance. Linear component measurements are also made using phasors by network analyzers and are compatible with harmonic balance.

The harmonic balance equations are formulated by insisting that Kirchoff's laws are satisfied at each frequency being considered. When evaluating nonlinear devices, it is often necessary to transform signals into and out of the time domain, but this is not essential and has no effect on the result, and so harmonic balance is considered a frequency-domain method.

The number of equations and unknowns in a harmonic balance formulation roughly equals two times the number of nodes times the number of frequencies considered. The factor of two results from needing both magnitude and phase in the phasor representation of a signal. When Newton-Raphson is used to solve these equations, the Jacobian gets very large. Thus memory and computation time requirements of harmonic balance can be considerable for large circuits when many frequencies are being considered. The Jacobian is sparse, and this fact must be exploited for harmonic balance to be practical for even moderately sized circuits. It is possible to further exploit sparsity by using the harmonic relaxation-Newton algorithm. Independent of which method is used to solve the nonlinear harmonic balance equations, harmonic balance is well suited to implementation on parallel and vector processors.

If harmonic balance were to be converted into a time-domain method, it would be a finite-difference method. By starting from (5.5), harmonic balance is written as a finite-difference method by simply multiplying through by the inverse almost-periodic Fourier transform matrix Γ^{-1}. For node n,

$$\Gamma^{-1}I_n(V) + \Gamma^{-1}\Omega_{nn}Q_n(V) + \Gamma^{-1}\sum_{m=1}^{N}Y_{mn}V_n + \Gamma^{-1}U_n = 0.$$

Knowing that $V_n = \Gamma v_n$, $I_n(V) = \Gamma i_n(v)$, $Q_n(V) = \Gamma q_n(v)$ and $U_n = \Gamma u_n$ allows us to simplify this to

$$i_n(v) + \Gamma^{-1}\Omega_{nn}\Gamma q_n(v) + \Gamma^{-1}\sum_{m=1}^{N}Y_{mn}\Gamma v_n + u_n = 0. \qquad (8.1)$$

$\Gamma^{-1}\Omega_{nn}\Gamma$ and $\Gamma^{-1}Y_{mn}\Gamma$ are constants and so can be precomputed for efficiency. Equation (8.1) shows that harmonic balance can be formulated in the time domain as a finite-difference method. The basic difference between the two approaches is that the frequency-domain version represents the solution using the coefficients of the sinusoids and the finite-difference method represents them in sampled-data form. Though both methods give the same answer, the matrices in the finite-difference method are denser and so that approach will be less efficient.

4. Examples

Here we look back at the circuits presented in Chapter 2 as motivation. It is explained how the techniques presented in this book would be applied to these circuits.

4.1. Self-Biasing Amplifier

Self-biasing amplifiers, such as the one shown in Figure 2.1, are difficult for transient-based simulators because they exhibit a time constant that is very long compared to the period of a typical input frequency and it is very difficult to predict just how long any turn-on transient will last. Any of the steady-state methods are suitable for use on such circuits. If the circuit contains distributed devices or behaving near-linearly, harmonic balance should be used, otherwise, shooting methods are likely to be more efficient.

4.2. Mixers

The wide frequency range and the long time constants present in mixers cause problems for transient analysis. Mixers, by their very nature, generally behave very nonlinearly. The closer the nonlinearities are to acting like ideal switches, the better the conversion efficiency and noise performance of the mixer. While it is possible to simulate these circuits with harmonic balance, it is expensive because a large number of frequencies will be needed to accurately represent the signals. If the mixer contains distributed devices, which is common at microwave frequencies, then harmonic balance is the only alternative to transient analysis. If the mixer is purely lumped, then quasiperiodic shooting methods based on the MFT algorithm provide an attractive alternative, especially if the circuit is small.

4.3. Narrow-Band Amplifiers and Filters

These circuits are difficult for transient analysis because of their long settling times (if they are high-Q), and because of the widely spaced frequencies present in a two-tone intermodulation distortion test. For lumped filters, either harmonic balance or shooting methods provide an attractive alternative to transient analysis. For a two-tone test, harmonic balance is particularly attractive because the circuit is generally behaving near linearly and so harmonic balance is efficient. For distributed filters, harmonic balance is the clear choice.

4.4. Low-Distortion Amplifiers

When driven with a sinusoid, low-distortion amplifiers respond in steady state with almost a pure single-tone sine-wave. The amount that the response differs from a pure sine-wave is a good measure of the distortion of the amplifier. Highly linear amplifiers will produce very small distortion products. Transient simulators use piecewise polynomials to represent the solution and so time-points must be placed close together so that the piecewise polynomial fits the sinewave well. For example, if the distortion products of the amplifier are 100 dB below the fundamental, then the time-points must be chosen close enough together so that piecewise polynomial fits a sine-wave to within 1 ppm in order to accurately resolve the distortion products. As the linearity of the amplifier is improved, the time-points must be chosen closer and closer together.

Harmonic balance gets more accurate as the level of the distortion products decrease without having to increase the number of frequencies computed. In fact, as the distortion goes to zero, harmonic balance becomes exact if the stimulus is representable exactly with the chosen number of harmonics. This is an important characteristic of harmonic balance that sets it apart from other methods of solving differential equations. Thus, harmonic balance is clearly the best method for simulating low-distortion amplifiers.

4.5. Switched-Capacitor Filters

Switched-capacitor filters have a large repetitive clock signal whose period is generally much shorter than the duration of the interval of interest. The high frequencies combined with the long simulation interval result in an expensive transient analysis. Harmonic balance is inappropriate with these filters because of the large number of harmonics in the clock signal, which is always a pulse train. If a sinusoidal input is applied to the filter, the resulting signals are quasiperiodic, and so the only real alternative to transient analysis is MFT-based shooting methods. This is particularly true if two or more sinusoids are applied to the input of the filter, which would be the case if the intermodulation distortion of the filter were being measured.

4.6. Traveling-Wave Amplifiers

The difficulty with traveling-wave amplifiers is that they contain transmission lines. These amplifiers are used up to very high frequencies and so various nonideal effects such as dispersion and loss must be accounted for in the lines. The dispersion and loss present problems for transient analysis and the distributed devices make shooting methods unsuitable. The only appropriate approach is harmonic balance, which usually works well. Traveling-wave amplifiers often exhibit conditional stability (stable for small signals and large signals, but with a range of instability in between). Harmonic balance will have trouble in this situation and so pseudo-arc-length continuation is needed.

4.7. Measured Devices

Measured devices are characterized with S-parameters measured in the frequency domain. As yet, no transient analysis simulator has been able to use such data in a simulation. The only proven simulation technique for nonlinear circuits that contain measured devices is harmonic balance.

4.8. Crystal and Cavity Oscillators

Crystal and cavity oscillators are designed for very high-Q, which implies that these circuits have very long turn-on transients. This makes transient methods inappropriate, but as long as the circuit is lumped, any of the steady-state methods for autonomous systems work fine. With cavity oscillators, the resonator is generally considered to be distributed, and so harmonic balance is the best choice.

Chapter 9
Summary

This book presented several new algorithms for finding periodic and quasiperiodic responses of analog and microwave circuits in particular and systems of nonlinear ordinary- and integro-differential equations in general. The new algorithms were implemented in two new circuit simulation programs and used to simulate a variety of practical circuits. The main new results in this book are summarized below.

Harmonic balance was presented as a method that differed from traditional transient analysis in that it approximated the solution to a differential equation as a sum of sinusoids (a Fourier series) rather than a piecewise polynomial. It also used the Fourier coefficients (or phasors) to represent the signals rather than a sampled version of the waveforms. In other words, the circuit simulation problem was formulated by writing Kirchoff's laws in the frequency domain. Using phasors to represent the signals provides important advantages when formulating and evaluating linear device models, particularly when the models are distributed. This is the main reason why harmonic balance is quickly becoming the dominant method for simulating nonlinear circuits at microwave frequencies. In general, nonlinear devices cannot be directly evaluated in the frequency domain, and so signals are converted between the frequency- and time-domains with a discrete Fourier transform. The nonlinear devices are evaluated in the time domain, with the results transformed back into the frequency domain. The DFT and APFT are used for transforming periodic and quasiperiodic signals, respectively. It was also shown that it is possible to avoid the APFT altogether and use the DFT even on quasiperiodic circuits. Doing so results in a noticeable improvement in speed.

Methods for solving the nonlinear system of equations formulated by harmonic balance were grouped into three broad categories, nonlinear programming techniques, nonlinear relaxation techniques, and the Newton-Raphson algorithm. A new method referred to as Gauss-Jacobi harmonic relaxation was shown to be the best of the relaxation methods, by virtue of it being fast and the most likely to converge. A new hybrid adaptive method based on Gauss-Jacobi harmonic relaxation and harmonic Newton was proposed. This method provides most of the speed of the Gauss-Jacobi relaxation method and the large region of convergence of the harmonic Newton method. The hybrid method, referred to as harmonic relaxation-Newton, was shown to be just as robust as harmonic Newton, but considerably faster. Harmonic relaxation-Newton was further accelerated by employing Šamanskii's method.

The Gauss-Jacobi harmonic relaxation, harmonic Newton, and harmonic relaxation-Newton methods were implemented in *Spectre* and the algorithms compared on many practical circuits. The results obtained indicate that the harmonic relaxation-Newton method provides the best tradeoff between speed and region of convergence. However, the hueristics used are immature. Considerable improvement in speed could be realized by carefully reworking the block matrix ordering algorithms and the algorithm that sets the block bandwidth.

A second algorithm presented for simulating quasiperiodic circuits is the mixed frequency-time (MFT) method. MFT maps the problem of finding the quasiperiodic solution of an ordinary differential equation approximately into a boundary-value problem. In the development of the algorithm, one of the fundamental frequencies of the quasiperiodic signals was chosen to be the clock. All signals were sampled at the beginning of each clock period. MFT's fundamental assumption was that the sampled waveforms, which were also shown to be quasiperiodic, could be accurately represented with a Fourier series with a small number of nonzero Fourier coefficients. Given that J Fourier coefficients were sufficient, then knowing J points on the sampled waveforms would be enough to allow the Fourier coefficients to be computed. From the coefficients, all points on the sampled waveforms can be found. In particular, if J points were known, then their adjacent successors could be computed solely from the assumption that sampled waveforms were accurately approximated with a J coefficient Fourier series. The relationship of the J points to their immediate successors through the truncated Fourier series assumption is referred to as the delay operator relation. Another relationship exist between the J points and their immediate successors through the state-transition function of the differential equation. These two relationships can be combined into one nonlinear system of equations that can be solved for the values of the signals at the J points. From there, the Fourier coefficients are computed, and then any of the points on the sampled waveform. If the solution is needed at some other time, the state-transition function can be evaluated starting at the closest preceding sample.

The MFT algorithm was implemented in *Nitswit* and was tested on a wide variety of lumped circuits, including switched-capacitor filters and switching mixers. It was shown to be practical on moderately sized circuits.

Recent related work was also presented in Appendix B, where a transform suitable for almost-periodic signals was derived from the matrix form of the DFT. Conceptually, it is easy to extend the matrix form of the DFT for almost-periodic signals, however, a naive implementation is likely to be inaccurate. The concepts of truncation and aliasing were introduced. Truncation is the operation of eliminating all but a finite number of frequencies from consideration. Aliasing refers to the amplification and conversion of those signals at the truncated frequencies into signals at frequencies of interest by the transform. The condition number of the transform matrix was shown to be a measure of the degree to which the aliased terms are amplified. For the transform to be accurate, it is important for the condition number to be as small as possible. A small condition number was shown to be achievable by choosing time-points for the transform so that the rows of the transform matrix are nearly orthogonal. A practical algorithm was given that chooses such a set of time-points. The combination of the almost-periodic transform matrix and the time-step algorithm was referred to as the APFT.

This book presented several new approaches to finding the periodic and quasiperiodic solutions of nonlinear differential equations, and new ways of improving the efficiency of the standard approaches. These results will eventually provide the designers of analog and microwave circuits useful new capabilities for their circuit simulators, and allow them to explore their designs more fully. However, there is certainly more work that can be done to further improve the efficiency of these approaches, and perhaps a breakthrough or two that remains to be uncovered. Also, efficient methods for determining whether these solutions are stable, and so represent steady-state solutions, are an important research topic that has not been adequately explored. Furthermore, except for standard transient analysis, no method exists today to compute a chaotic steady-state solution. This has become a very important problem with the growing popularity of delta-sigma modulators.

Appendix A
Nomenclature

$\mathbf{Z}$, $\mathbb{R}$, $\mathbf{C}$ The integer, real, and complex numbers.

$\mathbf{C} = \mathbb{R}^2$ Throughout most of this book, the trigonometric Fourier series is used rather than the exponential. Thus, a Fourier coefficient is described using the coefficients of sine and cosine. The pair of these two coefficients are said to reside in $\mathbf{C}$ as opposed to $\mathbb{C}$. $\mathbf{C}$ is related to $\mathbb{C}$ in that $[a,b]^{\mathrm{T}} \in \mathbf{C}$ corresponds to $a + \mathbf{j}b \in \mathbb{C}$.

$\| \cdot \|_\infty$ The l_∞ norm. For $x \in \mathbb{R}^N$, $\| x \|_\infty = \max_i |x_i|$. For $A \in \mathbb{R}^{N \times N}$, $\| A \|_\infty = \max_i \sum_{j=1}^N |a_{ij}|$.

$\| \cdot \|_2$ The Euclidean or l_2 norm. For $x \in \mathbb{R}^N$, $\| x \|_2 = \left[\sum_{i=1}^N x_i^2 \right]^{1/2}$. For vectors in $\mathbb{R}^N$, the l_2 and l_∞ norms are equivalent. That is, $\frac{1}{\sqrt{N}} \| x \|_2 \leq \| x \|_\infty \leq \| x \|_2$ for all $x \in \mathbb{R}^N$.

$O(\cdot)$ $f(x) = O(g(x))$ when $x \rightarrow a$ implies that $|f(x)/g(x)|$ is bounded as $x \rightarrow a$ (a can be finite, $+\infty$, or $-\infty$).

$\mathbf{j}$ Imaginary operator, $\mathbf{j} = \sqrt{-1}$.

0,1   The zero vector or matrix and the identity matrix.

t, ω Time, radial frequency.

λ A fundamental frequency.

Λ, Λ_K A countable set of frequencies, and a finite set with K elements.

$Z(T;\theta)$ A sequence of equally spaced points. The distance between points is T and they are offset from zero by θ. $Z(T,\theta) = \{t : t = sT + \theta, s \in \mathbf{Z}\}$.

$P(T;E)$ The space of all periodic waveforms of bounded variation with period T with domain E.

$AP(\Lambda; E)$ The space of almost-periodic functions on domain E constructed as a linear combination of sinusoids at frequencies in the set Λ.

$QP(\lambda_1, \cdots, \lambda_d; E)$
                     The set of quasiperiodic functions on domain E with fundamental frequencies $\lambda_1, \lambda_2, \cdots, \lambda_d$. Equals $AP(\Lambda; E)$ where Λ is the module constructed from the basis of fundamental frequencies.

$\mathbb{F}, \mathbb{F}^{-1}$ Abstract forward and inverse Fourier operators.

Γ, Γ^{-1} Matrix representation of the forward and inverse Fourier operators.

x, X Arbitrary waveform and its spectrum. $X = \mathbb{F}x$.

$\longleftrightarrow$ Laplace transform relation.

f Function that maps waveforms to waveforms. Sometime f is an arbitrary differentiable function, other times it is used to represent the sum of currents entering a node or nodes.

$J_f(x_0)$ Jacobian (derivative) of f with respect to x at x_0.

F Function that maps spectra to spectra. Related to f in that if $y = f(x)$ then $Y = F(X)$.

H	The maximum number of harmonics considered.
K	The number of frequencies present in the spectra.
N	The number of nodes in a circuit.
S	The number of time-points present in the sampled waveforms.
k, l	Frequency indices. Usually, $k, l \in \{0, 1, \cdots, K-1\}$
m, n	Node indices. $m, n \in \{1, 2, \cdots, N\}$
r, s	Time indices. $r, s \in \{0, 1, \cdots, S-1\}$
v, V	Node voltage waveforms, spectra.
u, U	Input current waveforms, spectra.
i, I	Function from voltage to current for nonlinear resistors and its frequency-domain equivalent.
q, Q	Function from voltage to charge for nonlinear capacitors and its frequency-domain equivalent.
y	Matrix-valued impulse response of the circuit with all nonlinear devices removed.
Y	Laplace transform of y.
$\mathcal{Y}$	Phasor equivalent to Y.
Ω	Matrix used to multiply each particular frequency component in a vector of spectra by the correct ω_k to perform the frequency-domain equivalent of time differentiation.
ϕ	The state transition function. For a differential equation $\phi(x_0, t_0, t)$ maps an initial state x_0 at time t_0 into the state $x(t)$ at time t.
Φ	The composite state transition function. For a differential equation $\Phi(X_0, t_0, t)$ maps a composite initial state X_0 at time t_0 into the composite state $X(t)$ at time t.
Γ_0, Γ_T	Γ_0 is a matrix that maps a waveform that has been sampled at a finite collection of time-points into its

spectrum. Γ_T is a matrix that maps the same waveform sampled at the same time-points delayed T seconds into its spectrum.

$D(T)$ The delay operator. $D(T)$ is a matrix that maps a waveform that has been sampled at a finite collection of time-points into the same waveform sampled at the same time-points delayed T seconds. $D(T) = \Gamma_T^{-1}\Gamma_0$.

$D_N(T)$ The composite delay operator. $D_N(T)$ is a matrix that maps a composite waveform that has been sampled at a finite collection of time-points into the same waveform sampled at the same time-points delayed T seconds.

h_s The time step. $h_s = t_s - t_{s-1}$.

Appendix B
APFT Time-Point Selection

This chapter describes the APFT in detail. It includes a discussion of error and the description of a nonuniform time-point selection algorithm.

1. Matrix Formulation

To make the Fourier operator computationally tractable, it is necessary to consider only a finite number of frequencies. Denote the finite set of frequencies as $\Lambda_K = \{\omega_0, \omega_1, \omega_2, \ldots, \omega_{k-1}\}$ and assume all frequencies are distinct ($\omega_j \neq \omega_k$ when $j \neq k$) and that $\omega_0 = 0$. By considering only a finite number of frequencies, it is possible to sample a waveform at a finite number of time-points and calculate its Fourier coefficients. Since the spaces involved are now finite dimensional, the first representation theorem of linear algebra shows that the Fourier transform $\mathbb{F}$ and its inverse $\mathbb{F}^{-1}$ can be viewed as matrices acting on the vectors of samples and coefficients, respectively. That is,

$$\sum_{\omega_k \in \Lambda_K} (X^C(k)\cos\omega_k t + X^S(k)\sin\omega_k t) = x(t)$$

can be sampled at S time-points, resulting in the set of S equations and $2K - 1$ unknowns

$$
\Gamma^{-1}
\begin{bmatrix}
X(0) \\
X^C(k) \\
X^S(k) \\
\vdots \\
X^C(K-1) \\
X^S(K-1)
\end{bmatrix}
=
\begin{bmatrix}
x(t_1) \\
x(t_2) \\
x(t_3) \\
\vdots \\
x(t_S)
\end{bmatrix}
\tag{B.1 a}
$$

where

$$
\Gamma^{-1} =
\begin{bmatrix}
1 & \cos\omega_1 t_1 & \sin\omega_1 t_1 & \cdots & \cos\omega_{K-1} t_1 & \sin\omega_{K-1} t_1 \\
1 & \cos\omega_1 t_2 & \sin\omega_1 t_2 & \cdots & \cos\omega_{K-1} t_2 & \sin\omega_{K-1} t_2 \\
1 & \cos\omega_1 t_3 & \sin\omega_1 t_3 & \cdots & \cos\omega_{K-1} t_3 & \sin\omega_{K-1} t_3 \\
\vdots & \vdots & \vdots & & \vdots & \vdots \\
1 & \cos\omega_1 t_S & \sin\omega_1 t_S & \cdots & \cos\omega_{K-1} t_S & \sin\omega_{K-1} t_S
\end{bmatrix}
\tag{B.1 b}
$$

If the frequencies ω_k are distinct, and if $S = 2K - 1$, this system is invertible for almost all choices of time-points, and can be compactly written as $\Gamma^{-1} X = x$. Inverting Γ^{-1} gives $\Gamma x = X$. Γ and Γ^{-1} are a discrete Fourier transform pair.

Given a finite set Λ_K of distinct frequencies, and a set of time-points, we say that Γ and Γ^{-1} are one implementation of the almost-periodic Fourier transform for $AP(\Lambda_K; \mathbb{R})$. Once Γ and Γ^{-1} are known, performing either the forward (using Γ) or inverse (using Γ^{-1}) transform requires just a matrix multiply, or $(2K - 1)^2$ operations; this is the same number of operations required by the DFT.

The DFT is a special case of (B.1) with $\omega_k = k\omega$ ($k = 0, 1, 2, \ldots, K-1$) and $t_s = sT/S$ ($s = 1, 2, \ldots, S$), i.e. when the frequencies are all multiples of a single fundamental and the time-points are chosen equally spaced within the period. The DFT and its inverse, the IDFT, have the desirable property of being well conditioned, which is to say that very little error is generated when transforming between x and X. From the matrix viewpoint, the high accuracy of the DFT corresponds to the fact that the rows of Γ^{-1} are orthogonal. (More is said about this later.) Unfortunately, the DFT and the IDFT are defined only for periodic signals.

For almost-periodic signals, if the time-points are not chosen carefully, Γ^{-1} can be ill-conditioned. In particular, choosing time-points to be equally spaced often is a bad strategy when signals are not periodic. Unlike the periodic case, it is in general impossible to choose a set of time-points over which the sampled sinusoids at frequencies in Λ_K are orthogonal. In fact, it is common for evenly sampled sinusoids at two or more frequencies to be nearly linearly dependent, which causes the ill-conditioning problems encountered in practice. This ill-conditioning can greatly magnify aliasing. Thus, it is important to choose a set of time-points that results in well-conditioned transform matrices.

1.1. Previous Work

Ushida and Chua [ushida84] use equally spaced time-points, but avoid the ill-conditioning problem by using extra time-points. In doing so, the matrix Γ^{-1} becomes a tall rectangular matrix. To make the system square again, both sides of (B.1) are multiplied by $(\Gamma^{-1})^T$, which results in

$$(\Gamma^{-1})^T\Gamma^{-1}X = (\Gamma^{-1})^Tx.$$

Thus (B.1) is converted into a least squares problem that is solved in the traditional manner using the normal equation. Unfortunately, the normal equation is notoriously ill-conditioned and so a new ill-conditioning problem may be introduced.

Gilmore [gilmore86] samples the waveform using several small sets of equally spaced time-points. The DFT is applied to each set individually. The sets are too small to prevent aliasing in the computed spectra. The aliasing is eliminated by taking an appropriate linear combination of the computed spectra. Since the DFT is used, the method is constrained to periodic signals, though it can be much more efficient than the standard DFT on sparse spectra. The total number of time-points used is normally greater than the theoretical minimum by about 50%. The numerical stability of this approach is unknown.

1.2. Condition Number and Orthonormality

It is now necessary to discuss the conditioning of a system of equations, a concept alluded to earlier. Formally, the condition number of a matrix A is defined as $\kappa(A) = \| A \| \ \| A^{-1} \|$ [golub83]. The condition number of a matrix is important because it is a measure of how much errors can be amplified during the course of solving a matrix equation. For example, consider solving $Ax = b$ for x when both A and b are contaminated with error. Write the contaminated system as

$$(A + \delta A)(x + \delta x) = b + \delta b.$$

If $\| \delta A \|$ and $\| \delta b \|$ are small, then $\| \delta x \|$ can be bounded [golub83] with

$$\frac{\| \delta x \|}{\| x \|} \leq \kappa(A) \left[\frac{\| \delta A \|}{\| A \|} + \frac{\| \delta b \|}{\| b \|} \right] + higher\ order\ terms$$

The problem of ill-conditioning in (B.1) can be visualized by considering each equation as defining a hyperplane in the Euclidean space $\mathbb{R}^{2K-1}$. Let $\rho_s \in \mathbb{R}^{2K-1}$ be such that ρ_s^T is the sth row in Γ^{-1}; then the sth hyperplane is defined as the set of all points X such that $\rho_s^T X = x(t_s)$. Thus, ρ_s is a vector orthogonal to the hyperplane. The solution to (B.1) is the intersection of all the hyperplanes. If the system is degenerate because two or more planes are coincident, then the intersection is not a single point and the system of equations has an infinite number of solutions. If there are no coincident hyperplanes, but two or more of the planes are nearly parallel, then a unique solution exists; however, high-precision arithmetic is needed to find it accurately.

A matrix is degenerate if and only if there is a linear dependence among its row vectors, and it is natural to suppose that a matrix has a small (good) condition number if its rows are nearly orthonormal (and thus "far" from being linearly dependent). We now prove this to be true.

Consider an invertible $N \times N$ matrix A. Suppose that the rows a_n of A, regarded as vectors, are nearly orthonormal. In particular, suppose that each vector has unit Euclidean length and that the orthogonal component of each vector a_n with respect to the space S_n spanned by the others is at least $\alpha \leq 1$ (it would be exactly 1 if the vectors were precisely orthonormal).

When forming the product $A^{-1}A = 1$, each row of A^{-1} can be thought of as the coefficients of a linear combination of the rows of A. This linear combination yields a row in the identity matrix — a vector of length 1. Suppose that the nth element in a row of A^{-1} has absolute value $r > 1/\alpha$. Then the component of the resulting linear combination that is in the direction orthogonal to S_n is determined solely by $r a_n$, and will have magnitude greater than $r \alpha > 1$. Since the linear combination is a vector of unit length, this is a contradiction. Therefore, no element of any row of A^{-1}, and thus no element of A^{-1}, has absolute value greater than $1/\alpha$.

Since $A \in \mathbb{R}^{N \times N}$, it follows that $\| A^{-1} \|_\infty$ (the l_∞ norm of A^{-1}) is no more than N/α. And since, by assumption, the Euclidean norm of the rows of A equals one, $\| A \|_\infty \leq N$ (employing the equivalence of the l_2 and l_∞ norms in $\mathbb{R}^N$), and therefore, $\kappa(A) \leq N^2/\alpha$. In short, the near-orthonormality of a matrix places an upper bound on its condition number.

Note that multiplying a matrix by a scalar β does not affect its condition number, since the norms of the matrix and its inverse are respectively multiplied by β and $1/\beta$. Thus, if all rows of a matrix have equal Euclidean length (not necessarily one) and, when scaled to one, satisfy the orthonormality property, the matrix is still well-conditioned. If the rows of a matrix are nearly orthonormal after they have been scaled to unit length, we say that they are (or the original matrix is) nearly orthogonal.

1.3. Condition Number and Time-Point Selection

Given a finite set of frequencies Λ_K, any set of $S = 2K - 1$ time-points yields a Γ^{-1} whose row vectors (consisting of a single 1 and a set of sine-cosine pairs) have Euclidean norms $\sqrt{K}$. Thus, if we could find a set of time-points so that these rows were nearly orthogonal, it would follow from the discussion above that Γ^{-1}, and therefore Γ would be well conditioned.

However the relation between the time-points and the orthogonality of the resultant row vectors is clearly rather involved; finding a set of times which define nearly-orthogonal row vectors seems to be quite difficult. One approach is to write down *a priori* a set of orthogonal vectors and then look for time-points that generate vectors close to these pre-specified ones; this is equivalent to defining the approximate phases of each sine wave and looking for a time where every wave is in the appropriate phase. This in turn can be thought of as a set of approximate equalities modulo 2π, but it is far from clear under what circumstances a solution exists or how to go about finding it.

Another approach is to choose time-points equally spaced within a time interval larger than the period corresponding to the smallest nonzero frequency in Λ_K. As we discuss later however, experience shows that this method of time-point selection is unsatisfactory.

1.4. Condition Number and Aliasing

As mentioned previously, the condition number provides a measure of how much the error is amplified during a calculation. Roundoff is one source of error in the transform, but there is another that is normally much larger — the error due to truncating Λ to Λ_K. This error consists of two pieces. The Fourier coefficients of the frequencies omitted from Λ are presumably small but may not be exactly zero. Neglecting these frequencies is referred to as truncation error. While neglected in the transforms Γ and Γ^{-1}, these frequencies still exist in the vector of samples x. They masquerade as frequencies in Λ_K and result in further error that is referred to as aliasing. Because of these errors, the computation of X will be in error.

Fortunately, this error can be bounded. Suppose that the overlooked sinusoids contribute an error δx to the observed sample vector $x + \delta x$. From this we calculate the Fourier coefficients $X + \delta X$ using

$$X + \delta X = \Gamma(x + \delta x).$$

By construction we know that $X = \Gamma x$. Thus, $\delta X = \Gamma \delta x$, and $\| \delta X \| \leq \| \Gamma \| \ \| \delta x \|$. By definition, $\kappa = \| \Gamma \| \ \| \Gamma^{-1} \|$. It is easily shown that $K \leq \| \Gamma^{-1} \|_{\infty} < \sqrt{2}K$, so $\| \Gamma \|_{\infty} \leq \kappa_{\infty}/K$ and

$$\| \delta X \|_{\infty} \leq \frac{\kappa_{\infty}}{K} \| \delta x \|_{\infty} . \tag{B.2}$$

That is, κ_{∞}/K is the upper bound on how much the error due to coefficients of truncated frequencies is amplified in the process of transforming a waveform to the frequency domain. In practice, error amplification factors often approach this bound, so it is very important to select a set of time-points such that κ is small.

The condition number gives a bound on the error that results from aliasing, but it gives little insight and can be pessimistic. Perhaps a better approach to exploring the errors in the APFT is to view the transform as a collection of K filters, each of which is responsible for computing one Fourier coefficient[1]. These filters take as input a sequence of S samples and output one coefficient. Such filters are referred to as Finite Impulse Response (FIR) filters [rabiner78] [rabiner75]. Consider the particular filter that computes the coefficient $X(j)$ from $\{x_s\}_{s=0}^{S-1}$. If $x_s = X^C(k)\cos\omega_k t_s + X^S(k)\sin\omega_k t_s$, $s = 0, 1, \ldots, S - 1$, then X_j must equal zero if $j \neq k$ and must equal $[X^C(k), X^S(k)]$ if $j = k$. In other words, the filter must have zeros in its transfer response at each frequency $\omega_k \neq \omega_j$ when $\omega_k \in \Lambda_K$, and unity gain when $\omega_k = \omega_j$. These constraints are satisfied by design. However, the question remains, what is the response of the filter when driven at a frequency $\omega \notin \Lambda_K$? In particular, what is the response when $\omega \notin \Lambda_K$ but $\omega \in \Lambda$? It is guaranteed by (B.2) that if $x_s = \cos(\omega t_s + \theta)$ then $\overset{\max}{\omega} |X| \leq \frac{\kappa_{\infty}}{K}$.

The frequency response of a typical Fourier coefficient filter is shown in Figure B.1. For this filter, $\omega_0, \omega_1, \ldots, \omega_5 \in \Lambda_K$,

---

[1]Edward Ngoya of Limoges University in France was the first to suggest this interesting and useful interpolation of the APFT.

$\omega_6, \omega_7, \ldots, \omega_9 \notin \Lambda_K$, and $\omega_0, \omega_1, \ldots, \omega_9 \in \Lambda$. The actual error due to aliasing is proportional to the response of the filter at frequencies in Λ but not in Λ_K. The condition number bound κ_∞/K may (or may not) be a pessimistic estimation of this error. So far there has been no attempt to choose time-points to minimize the filter response at frequencies in Λ but not in Λ_K. Choosing time-points in this manner would minimize the actual aliasing rather than the bound on the max-

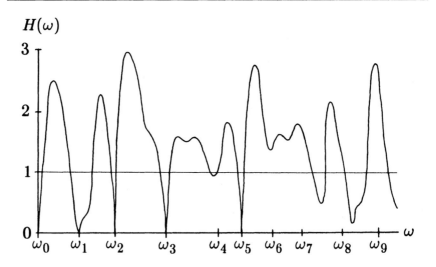

Figure B.1: Response of a typical APFT coefficient filter.

imum aliasing.

2. Near-Orthogonal Selection Algorithm
2.1. Time-Point Selection

Our time-point selection algorithm, referred to as *near-orthogonal selection*, was conceived using some of the ideas discussed above.

First, we thought that if selecting evenly-spaced time-points was likely to yield row vectors particularly close to being linearly dependent, we might be better off selecting time-points randomly from a time interval larger than the period corresponding to the smallest nonzero frequency in Λ_K. (We chose an interval equal to three times this period.) Such a choice is particularly attractive given the complexity of the relationship between the time-points and the orthogonality of the row vectors; making any more intelligent choice of time-points seem quite difficult.

Second, we realized that in essence the problem in recovering X from x is that the linear system may be close to being underdetermined, in a numerical sense. So adding additional equations should increase the accuracy of the calculation of X. In fact, if more than S time-points are chosen, Γ^{-1} becomes a tall rectangular matrix, and its pseudo-inverse Γ is a wide rectangular matrix satisfying $X = \Gamma x$.

Oversampling with twice as many randomly-selected time-points as theoretically necessary proves to be successful: it yields a very well conditioned system. However, when using the transform in the context of harmonic balance, all the nonlinear devices must be evaluated at each time-point. This is expensive because of the complexity of the nonlinear device models and because of the additional operations needed when forming the conversion matrices. Thus, oversampling is a costly remedy. It is clear, however, that the rows of the tall Γ^{-1} matrix span the space well (in a numerical sense). Perhaps some carefully chosen subset of these rows might also suffice.

The near-orthogonal selection algorithm takes just this approach; from a Γ^{-1} whose dimension is M rows by S columns, where $S = 2K - 1$ and $M > S$, it selects a set of just S rows, thus requiring no extra time samples. In other words, from a pool of more row candidates than necessary (we chose $M = 2S$, which seems to give good results in practice) and their corresponding time-points, a "good" minimal set is selected during the initialization of the algorithm. When actually performing the transform, only the minimal set of time-points is used. With harmonic balance, all nonlinear devices are evaluated at each time-point. That only the minimum number of time-points is used, and not 1.5 to 2 times the minimum as required by the other methods, is one of the significant advantages of the APFT algorithm.

The near-orthogonal selection algorithm is a variation of the Gram-Schmidt orthogonalization procedure [dahlquist74]. Its input is the matrix formed by randomly choosing twice as many time-points as necessary and forming the corresponding row vectors, ρ_s. Initially, these vectors all have the same Euclidean length (i.e., l_2 norm). One of these vectors, say ρ_1, is chosen arbitrarily. Any component in the direction of ρ_1 is removed from the remaining vectors using

$$\rho_s \leftarrow \rho_s - \frac{\rho_1^T \rho_s}{\rho_1^T \rho_1} \rho_1 \qquad\qquad s = 2, \cdots, M. \qquad (B.3)$$

The vectors that remain are now orthogonal to ρ_1. Since the vectors initially had the same length, the largest remaining vector was originally most orthogonal to ρ_1. It is chosen to play the role of ρ_1 for the next iteration of the algorithm. This process repeats until the required S vectors have been chosen. The time-points that correspond to these vectors are the time-points used to form Γ^{-1}. This algorithm is detailed below.

APFT Near-Orthogonal Selection Algorithm

Given:

$\Lambda_K = \{0, \omega_1, \omega_2, \cdots, \omega_{K-1}\}$, the set of frequencies.

Task:

To find a set of $S = 2K - 1$ time-points that results in a well-conditioned Γ^{-1}.

Algorithm:

$\omega_{min} \leftarrow \min(\{|\omega_k| : 1 \le k < K\})$

for $(s \leftarrow 1, \ldots, M)$

{   **random()** returns numbers uniformly distributed between 0 and 1.

$t_s \leftarrow \dfrac{6\pi}{\omega_{min}}$**random()**

$\rho_s^{(1)} \leftarrow [\, 1 \quad \cos\omega_1 t_s \quad \sin\omega_1 t_s \,, \ldots, \quad \cos\omega_{K-1} t_s \quad \sin\omega_{K-1} t_s \,]^T$

}

for $(r \leftarrow 1, \cdots, S)$

{   **argmax()** returns the index of the largest member of a set.

$k = $ **argmax**$(\{\, \| \rho_s^{(r)} \| \; : r \le s \le M \})$

swap$(\rho_r^{(1)}, \rho_k^{(1)})$

swap$(\rho_r^{(r)}, \rho_k^{(r)})$

swap(t_r, t_k)

for $(s \leftarrow r, \cdots, M)$

$$\rho_s^{(r+1)} \leftarrow \rho_s^{(r)} - \frac{\rho_r^{(r)T}\rho_s^{(r)}}{\rho_r^{(r)T}\rho_r^{(r)}}\rho_r^{(r)}$$

}

Results:

The set $\{t_s : 1 \le s \le S\}$ contains the desired time-points.

Once the time-points are selected, Γ^{-1} is constructed with the rows $\rho_s^{(1)}$ for $s = 1, \cdots, S$. It is easy to verify that the time-points are well-chosen by calculating the condition number

$$\kappa = \| \Gamma \| \ \| \Gamma^{-1} \| .$$

2.2. Constructing the Transform Matrix

There is another problem that up to now we have ignored. The arguments to the sine and cosine functions in (B.1) are potentially very large, which results in excessive roundoff error. For example, assume $\lambda_1 = 2\pi 10^9$ and $\lambda_2 = 2\pi(10^9 + \sqrt{2})$. Then $\omega_{\min} = 2\pi\sqrt{2}$ and so the time-points fall between 0 and $3/\sqrt{2}$ seconds. Thus, $\omega_l t_s$ can be as large as 10^{11}, causing two problems. First, on most computer systems, the trigonometry routines are not designed to handle such large arguments and often return meaningless results. This problem is easily avoided by subtracting from the argument as many multiples of 2π as possible without making it negative. The second problem is more troublesome. The approximately 10^{10} multiples of 2π in the argument have no effect on the result except to reduce its accuracy by about 10 digits. Since the $\omega_l t_s$ product must be formed (and so truncated to a finite number of digits by the computer) before the multiples of 2π can be removed, the digits are lost and cannot be reclaimed. While this error cannot be eliminated, it can be controlled by assuming Λ_K is a truncated module (note that up to now we have placed no restrictions on the frequencies in Λ_K except that they be distinct and that $\omega_0 = 0$). From (3.6), the product $\omega_l t_s$ can be written

$$\omega_l t_s = \sum_{j=1}^{d} k_j \lambda_j t_s .$$

Let

$$\psi_{js} = \textbf{fract}\left[\frac{\lambda_j t_s}{2\pi}\right] \qquad (B.4)$$

and

$$\phi_{ls} = 2\pi \sum_{j=1}^{d} k_j \psi_{js} \qquad (B.5)$$

where $1 \le j \le d$ and $1 \le s \le S$.

Now $\phi_{ls} = \omega_l t_s - 2\pi m$, where m is some integer and $|\phi_{ls}| \leq 2\pi \sum_{j=1}^{d} |k_j|$. Since the k_j are small integers, ϕ_{ls} is an appropriate argument to trigonometry routines on all computers. Because the product $t_s \lambda_j / 2\pi$ is formed before the **fract**() operator (which removes any integer portion and leaves only the fractional part) is applied, it is the dominant source of roundoff error. By using (B.4) and (B.5), the roundoff error can be viewed as resulting from roundoff error in the λ_j and t_s. Since the t_s are chosen randomly, their roundoff errors are of no concern. It was shown in Chapter 6 that when used with harmonic balance, even errors in λ_j are of no concern.

2.3. APFT Algorithm Results

The APFT near-orthogonal selection algorithm requires on the order of $M^2 S$ operations, where M is the number of time-point candidates used, and $S = 2K - 1$, where K is the number of frequencies. Since we have used $M = 2S$, the asymptotic complexity of the algorithm is the same order as that of the matrix inversion needed to compute Γ.

We note that while the initialization of the APFT (that is the time-point selection, formation of Γ^{-1}, and the inversion of Γ^{-1} to find Γ) requires on the order of S^3 operations, the actual forward and inverse transform requires S^2 operations, the same as the DFT. Thus, the expensive part of the APFT is performed only once per set of frequencies; after this initial overhead has been paid, the APFT is as efficient as the DFT.

To show the numerical stability of our method, we compare the condition number of Γ^{-1} when time-points are 1) evenly spaced, 2) randomly spaced, and 3) determined by the near-orthogonal selection algorithm. The condition number κ is roughly proportional to the errors in computing the inverse. Bear in mind that even the DFT, which is theoretically the best conditioned algorithm for the simpler periodic case, has a condition number $\kappa \approx N$, so the best we can hope for is linear growth of the condition number with the number of Fourier coefficients. Observe that, as shown by the results given in Figure B.2, the condition number from near-orthogonal selection is experimentally observed to grow linearly with K. That of random selection appears to grow quadratically, and that of evenly spaced grows exponentially.

The example chosen for comparison was with two fundamentals $\lambda_1 = 2\pi 10^9$ and $\lambda_2 = 2\pi(10^9 + \sqrt{2})$. Thus, the fundamentals differ by only 1 part in 10^9; also, because the fundamentals are incommensurable, the signal is not periodic. Truncation was performed using (3.6). Comparisons of the condition numbers are shown in Figure B.2 with the order H varying between 1 and 10. To smooth the wide variation seen in the results for the case of randomly selected time-points, each condition number plotted is the geometric mean of 10 trials. Similarly,

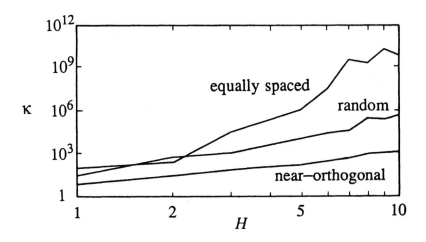

Figure B.2: Condition number of Γ^{-1} versus order H for the two fundamental APFT with truncation performed using (3.6) and time-points chosen evenly spaced, randomly, and using the near-orthogonal selection algorithm.

because different intervals give widely varying results for evenly-spaced points, those condition numbers are geometrically averaged over 10 intervals ranging from 1.5 to 4.5 times $2\pi/\omega_{min}$. Results obtained from near-orthogonal selection are so consistent that no averaging was needed, as evidenced by the smoothness of that curve. Graphing the condition number clearly shows that both randomly chosen and equally spaced samples have accuracy problems when the number of frequencies is large. Near-orthogonal selection from $2S$ randomly selected time-points always results in a reasonable condition number. Table B.1 gives a summary of information on the APFT with the near-orthogonal selection algorithm. Execution times were measured using the C programming language on a VAX 8650 running ULTRIX 2.0.

Recall that coefficients of frequencies not in Λ_K can be amplified by up to κ/K. For order $H = 10$, this amplification factor equals approximately 10^8 for evenly spaced points, 2000 for randomly spaced points, and 10 for points chosen using near-orthogonal selection. Thus, even if the coefficients of neglected frequencies are small, for evenly and randomly spaced points, the error δX due to truncation may be so large as to dominate over the desired coefficients X.

APFT Summary					
H	K	S	κ	t_{init}	$t_{transform}$
1	3	5	6	17 ms	0
2	7	13	24	67 ms	0.3 ms
3	13	25	64	280 ms	1.7 ms
4	21	41	113	1.1 s	3.6 ms
5	31	61	143	3.3 s	8.5 ms
6	43	85	270	8.6 s	17 ms
7	57	113	420	20 s	30 ms
8	73	145	790	41 s	49 ms
9	91	181	950	79 s	77 ms
10	111	221	1200	142 s	116 ms

Table B.1: Error estimates and execution times for the APFT algorithm using double precision arithmetic on a VAX 8650 with $\lambda_1 = 10^9$, $\lambda_2 = 10^9 + \sqrt{2}$, and truncation performed using (3.6). H is the number of harmonics of each fundamental. K is the total number of frequencies and S is the number of time samples. κ is the condition number of Γ^{-1}. t_{init} is the time required to choose the time-points and form and invert Γ^{-1}. $t_{transform}$ is the time required to multiply either Γ^{-1} or Γ by a vector.

Appendix C
Arc-Length Continuation

In Chapter 3 we introduced Newton-Raphson as a method of solving systems of nonlinear algebraic equations. Newton-Raphson is not guaranteed to find a solution unless an initial guess that is close enough to a solution is provided. Continuation methods were presented as a way of generating these initial guesses. It was shown that continuation methods can fail if the path of solutions generated by these method contains a limit point.

One solution to the limit point problem is to use a *path following* or *arc-length* continuation method. These methods treat the manifold of solutions in $\mathbb{R}^N \times [0,1]$ of (3.33) as a path that must be followed from $(x_0, 0)$ to $(\hat{x}, 1)$ by, at each step, solving for a pair (x, p) that solves (3.33) and makes some progress along the path. The independent parameter becomes arc length or the distance traveled along the path. Rather than specifying p apriori, it is considered a function of the arc-length parameter α. The arc length α is specified and p is found as a function of α. In this way, p is allowed to increase or decrease as needed to provide adequate progress along the path at each step.

The derivation of the pseudo-arc-length continuation algorithm [bolstad86] begins by defining $\xi \in \mathbb{R}^N \times [0,1]$ as $\xi = [x, p]^T$ and $c = \{\xi : f(\xi) = 0\}$. Thus, c is the curve or manifold of all solutions to (3.33). For simplicity, assume that there is only one curve, that it is connected, and that it passes through $(s_0, 0)$ and $(\hat{x}, 0)$. This curve is parameterized using the arc length α from some arbitrary point. Rewriting (3.33) to show the explicit dependence of x and p on the arc-length parameter α.

$$f(x(\alpha), p(\alpha)) = 0 \qquad (C.1)$$

To simplify notation, the dependence of ξ, x, and p on α will be implied rather than denoted explicitly.

The first step is, given the solution at the current step, predict the value of the solution at the next step. Do this by extrapolating along the tangent to the path. Let ξ_0 the current solution and $u(\alpha_0)$ be the unit vector tangent to c at $\xi_0 = \xi(\alpha_0)$. Then $u = [u_x, u_p]^T$ is such that

$$\left[\frac{df(\xi)}{d\xi}\right]^T u = \left[\frac{df(x,p)}{dx}\right]u_x + \frac{df(x,p)}{dp}u_p = 0 \qquad (C.2a)$$

and

$$\| u_x \|_2^2 + u_p^2 = 1, \qquad (C.2b)$$

where $\| u_x \|_2$ is the Euclidean norm of u_x. The first equation states that u is tangent to the curve and the second that u has unit length. The quantities u_x and u_p can be computed by dividing both sides of (C.2a) by u_p.

$$\frac{df(x,p)}{dx}\frac{u_x}{u_p} + \frac{df(x,p)}{dp} = 0$$

Let $\phi = u_x/u_p$, and find ϕ by solving

$$\frac{df(x,p)}{dx}\phi = -\frac{df(x,p)}{dp} \qquad (C.3)$$

Then, u_x and u_p are computed from

$$u_p = \frac{1}{\sqrt{1 + \| \phi \|_2^2}} \qquad (C.4)$$

$$u_x = u_p \phi \qquad (C.5)$$

Given ξ_0 and $u(\alpha_0)$, a prediction of the new solution $\xi_1 = \xi(\alpha_1)$ is computed from

$$x_1^{predict} = x_0 + u_x(\alpha_0)\delta\alpha \qquad (C.6a)$$

$$p_1^{predict} = p_0 + u_p(\alpha_0)\delta\alpha \qquad (C.6b)$$

where $x_1^{predict}$ and $p_1^{predict}$ are the predicted values of x_1 and p_1 and $\delta\alpha = \alpha_1 - \alpha_0$.

The next step is to formulate the dependence of both x and p on the arc length. It is not important that the distance between ξ_0 and ξ_1 along the curve be exactly $\delta\alpha$ because ξ_1 is only one step along the path to our true goal ξ_s, and ξ_1 itself is of no value. Since the distance ξ_0 and ξ_1 need only be roughly equal $\delta\alpha$, a computationally simpler *pseudo-arc-length* constraint is employed [bolstad86]. The pseudo-arc-length algorithm requires the new solution to satisfy the equations

$$f(\xi_1) = 0 \qquad (C.7a)$$

$$g(\xi,\alpha_1) = <u(\alpha_0),\xi_1 - \xi_0> - (\alpha_1 - \alpha_0) = 0 \qquad (C.7b)$$

where $< , >$ represents the inner product. Equations (C.7b) forces the new solution ξ_1 to lie on a hyperplane perpendicular to the tangent vector $u(\alpha_0)$. The hyperplane intersects the tangent at a distance $|\delta\alpha|$ from $\xi(\alpha_0)$ as shown in Figure C.1.[1] Applying Newton-Raphson to find the solution of (C.7) gives

$$\begin{bmatrix} \dfrac{df(\xi_1^{(j)})}{dx_1} & \dfrac{df(\xi_1^{(j)})}{dp_1} \\[2em] \dfrac{dg(\xi_1^{(j)},\alpha_1)}{dx_1}^{\mathrm{T}} & \dfrac{dg(\xi_1^{(j)},\alpha_1)}{dp_1} \end{bmatrix} \begin{bmatrix} x_1^{(j+1)} - x_1^{(j)} \\[1em] p_1^{(j+1)} - p_1^{(j)} \end{bmatrix} = - \begin{bmatrix} f(\xi_1^{(j)}) \\[1em] g(\xi_1^{(j)},\alpha_1) \end{bmatrix}$$

$$(C.8)$$

---

[1]Another choice for the pseudo-arc-length constraint is: $g = \| \xi(\alpha_1) - \xi(\alpha_0) \|_2^2 - |\alpha_1 - \alpha_0| = 0$. This forces the solution to lie on a sphere of radius $\delta\alpha$ centered about $\xi(\alpha_0)$.

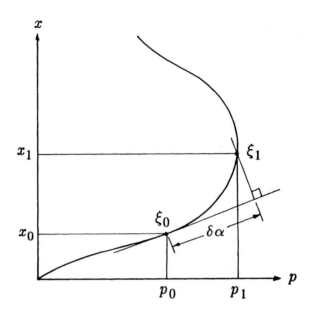

Figure C.1: Illustration of the pseudo-arc-length constraint.

At simple limit points the Jacobian in (C.8) is nonsingular [keller77], and so Newton-Raphson applied to the augmented system (C.7) is well-defined and quadratic convergence is possible.

If Gaussian elimination is used to solve (C.8), it is best to use pivoting to ensure numeric stability, especially near limit points. However, it is sometimes difficult to exploit the pattern of sparsity in the augmented system (this is true when the equations are generated by harmonic balance because df/dx is a block matrix while dg/dp is a scalar). Without pivoting, the algorithm has trouble moving over limit points because df/dx is singular at these points. In this situation, block Gaussian elimination without pivoting is used along with deflation techniques to avoid numerical instability [chan82, chan86]. The equations are now simplified further using block Gaussian elimination.

$$
\begin{bmatrix}
\dfrac{df\,(\xi_1^{(j)})}{dx_1} & \dfrac{df\,(\xi_1^{(j)})}{dp_1} \\[2ex]
0 & \dfrac{dg\,(\xi_1^{(j)},\alpha_1)}{dp_1} - \dfrac{dg\,(\xi_1^{(j)},\alpha_1)}{dx_1}^{T}\left[\dfrac{df\,(\xi_1^{(j)})}{dx_1}\right]^{-1}\dfrac{df\,(\xi_1^{(j)})}{dp_1}
\end{bmatrix} \times
$$

$$
\begin{bmatrix} x_1^{(j+1)} - x_1^{(j)} \\[1.5ex] p_1^{(j+1)} - p_1^{(j)} \end{bmatrix} = -
\begin{bmatrix} f\,(\xi_1^{(j)}) \\[2ex] g\,(\xi_1^{(j)},\alpha_1) - \dfrac{dg\,(\xi_1^{(j)},\alpha_1)}{dx_1}^{T}\left[\dfrac{df\,(\xi_1^{(j)})}{dx_1}\right]^{-1}\dfrac{df\,(\xi_1^{(j)})}{dp_1} \end{bmatrix}.
$$

Let

$$
\frac{df\,(\xi_1)}{dx_1}y = \frac{df\,(\xi_1)}{dp_1} \tag{C.9}
$$

so that

$$
y = \left[\frac{df\,(\xi_1)}{dx_1}\right]^{-1}\frac{df\,(\xi_1)}{dp_1}.
$$

And let

$$
\frac{df\,(\xi_1)}{dx_1}z = f\,(\xi_1) \tag{C.10}
$$

so that

$$
z = \left[\frac{df\,(\xi_1)}{dx_1}\right]^{-1}f\,(\xi_1).
$$

Then finally

$$\delta p^{(j)} = \frac{\dfrac{dg(\xi_1^{(j)},\alpha_1)}{dx_1^{\text{T}}}z - g(\xi_1^{(j)},\alpha_1)}{\dfrac{dg(\xi_1^{(j)},\alpha_1)}{dp_1} - \dfrac{dg(\xi_1^{(j)},\alpha_1)}{dx_1^{\text{T}}}y} \tag{C.11}$$

and

$$\delta x^{(j)} = -\left[\frac{df(\xi_1^{(j)})}{dx_1}\right]^{-1}\left[\frac{df(\xi_1^{(j)})}{dp_1}\delta p + f(\xi_1^{(j)})\right]$$

$$\delta x^{(j)} = -y\,\delta p^{(j)} - z. \tag{C.12}$$

The pseudo-arc-length continuation algorithm can be implemented using equations (C.3), (C.4), (C.5), (C.6), (C.7), (C.9), (C.10), (C.11), and (C.12). This algorithm is repeated at each new α_s, $s = 1, 2, \ldots, S$.

The pseudo-arc-length continuation method works if numerical pivoting is used when convergence is prevented by simple limit points (or rank one deficiencies in the original Jacobian) along the path. This is generally not the case when convergence problems occur during DC analysis, but is quite common when using harmonic balance on conditionally stable circuits.

Bibliography

[aprille72a]   Thomas J. Aprille and Timothy N. Trick. A computer algorithm to determine the steady-state response of nonlinear oscillators. *IEEE Transactions on Circuit Theory*, vol. CT-19, no. 4, July 1972, pp. 354-360.

[aprille72b]   Thomas J. Aprille and Timothy N. Trick. Steady-state analysis of nonlinear circuits with periodic inputs. *Proceedings of the IEEE*, vol. 60, no. 1, January 1972, pp. 108-114.

[baily69]   Everett Minnich Baily. *Steady-state harmonic analysis of nonlinear networks*. Ph. D. dissertation. Stanford University, February 1969.

[bava82]   G. P. Bava, S. Benedetto, E. Biglieri, F. Filicori, V. A. Monaco, C. Naldi, U. Pisani and V. Pozzolo. Modeling and performance simulation techniques of GaAs MESFET's for microwave power amplifiers. In *ESA-ESTEC REPORT*. Noordwijk, Holland, March 1982.

[bertsekas82]   Dimitri P. Bertsekas. *Constrained Optimization and Lagrange Multiplier Methods*. Academic Press, 1982.

[besicovitch32] A. S. Besicovitch. *Almost Periodic Functions*. Cambridge University Press, 1932. Also published by Dover 1954.

[bohr47]   Harald Bohr. *Almost Periodic Functions*. Chelsea, 1947.

[bolstad86]   John H. Bolstad and Herbert B. Keller. A multigrid continuation method for elliptic problems with folds.

SIAM Journal on Scientific and Statistical Computing, vol. 7, no. 4, October 1986, pp. 1081-1104.

[bracewell78] Ronald N. Bracewell. *The Fourier Transform and its Applications.* McGraw-Hill, 1978.

[brigham74] E. Oran Brigham. *The Fast Fourier Transform.* Prentice-Hall, 1974.

[chan82] Tony F. Chan. Deflation Techniques and Block-Elimination Algorithms for Solving Bordered Singular Systems. Technical Report #226, Department of Computer Science, Yale University, Yale Station, New Haven, CT 06520, March 10, 1982.

[chan86] Tony. F. Chan and Diana. C. Resasco. Generalized deflated block-elimination. *SIAM Journal on Numerical Analysis*, vol. 23, no. 5, October 1986, pp. 913-924.

[childs79] B. Childs, M. Scott, J. W. Daniel, E. Denman and P. Nelson (editors). *Codes for Boundary-Value Problems in Ordinary Differential Equations.* Springer-Verlag, 1979.

[chua75] Leon O. Chua and Pen-Min Lin. *Computer-Aided Analysis of Electronic Circuits: Algorithms and Computational Techniques.* Prentice-Hall, 1975.

[chua80] L. O. Chua. Device modeling via basic nonlinear circuit elements. *IEEE Transactions on Circuits and Systems*, vol. CAS-27, no. 11, November 1980, pp. 1014-1044.

[chua81] L. O. Chua and A. Ushida. Algorithms for computing almost periodic steady-state response of nonlinear systems to multiple input frequencies. *IEEE Transactions on Circuits and Systems*, vol. CAS-28, no. 10, October 1981, pp. 953-971.

[corduneanu68]   C. Corduneanu. *Almost Periodic Functions.* Interscience, 1968.

[cunningham58]   W. J. Cunningham. *Introduction to Nonlinear Analysis.* McGraw-Hill, 1958.

[curtice87]   Walter R. Curtice. Nonlinear analysis of GaAs MESFET amplifiers, mixers, and distributed amplifiers using the harmonic balance technique. *IEEE Transactions on Microwave Theory and Techniques,* vol. MTT-35, no. 4, April 1987, pp. 441-447.

[dahlquist74]   Germund Dahlquist and Åke Björck. *Numerical Methods.* Prentice-Hall, 1974.

[dalichow76]   Rolf Dalichow and Daniel Harkins. A precision RF source and down-converter for the model 8505a. *Hewlett-Packard Journal,* July 1976.

[de man80]   Hugo J. De Man, Jan Rabaey, Guido Arnout and Joos Vandewalle. Practical implementation of a general computer aided design technique for switched capacitor circuits. *IEEE Journal of Solid-State Circuits,* vol. SC-15, April 1980, pp. 190-200.

[desoer69]   Charles A. Desoer and Ernest S. Kuh. *Basic Circuit Theory.* McGraw-Hill, 1969.

[djordjevic86]   Antonije R. Djordjevic, Tapan K. Sarkar and Roger F. Harrington. Analysis of lossy transmission lines with arbitrary nonlinear terminal networks. *IEEE Transactions on Microwave Theory and Technique,* vol. MTT-34, no. 6, June 1986, pp. 660-666.

[duff86]   I. S. Duff, A. M. Erisman and J. K. Reid. *Direct Methods for Sparse Matrices.* Oxford University Press, 1986.

[egami74]   Shunichiro Egami. Nonlinear, linear analysis and computer-aided design of resistive mixers. *IEEE Transactions on Microwave Theory and Techniques,*

vol. MTT-22, no. 3, March 1974, pp. 270-275.

[engl82] W. L. Engl, R. Laur and H. K. Dirks. MEDUSA —
A simulator for modular circuits. *IEEE Transactions
on Computer-Aided Design of Integrated Circuits
and Systems*, vol. CAD-1, no. 2, April 1982, pp.
85-93.

[faber80] Maarek T. Faber and Wojciech K. Gwarek.
Nonlinear-linear analysis of microwave mixer with
any number of diodes. *IEEE Transactions on
Microwave Theory and Techniques*, vol. MTT-28,
no. 11, November 1980, pp. 1174-1181.

[fang83] S. C. Fang and et. al. Switcap: A switched-capacitor
network analysis program. *IEEE Circuits and
Systems Magazine*, vol. 5, no. 3, September 1983,
pp. 4-10, 41-46.

[filicori79] F. Filicori, V. A. Monaco and C. Naldi. Simulation
and design of microwave class-C amplifiers through
harmonic analysis. *IEEE Transactions on
Microwave Theory and Techniques*, vol. MTT-27,
no. 12, December 1979, pp. 1043-1051.

[fox57] L. Fox. *The Numerical Solution of Two-Point
Boundary Value Problems in Ordinary Differential
Equations*. Oxford, 1957.

[gautschi61] W. Gautschi. Numerical integration of ordinary
differential equations based on trigonometric
polynomials. *Numerische Mathemetik*, vol. 3, 1961,
pp. 381-397.

[gear71] C. William Gear. *Numerical Initial Value Problems
in Ordinary Differential Equations*. Prentice-Hall,
1971.

[gill81] Philip E. Gill, Walter Murray and Margaret H.
Wright. *Practical Optimization*. Academic Press,
1981.

[gilmore84]   R. J. Gilmore and F. J. Rosenbaum. Modeling of nonlinear distortion in GaAs MESFETs. *1984 IEEE MTT-S International Microwave Symposium Digest*, May 1984, pp. 430-431.

[gilmore86]   Rowan Gilmore. Nonlinear circuit design using the modified harmonic balance algorithm. *IEEE Transactions on Microwave Theory and Techniques*, vol. MTT-34, no. 12, December 1986, pp. 1294-1307.

[golub83]   Gene H. Golub and Charles F. Van Loan. *Matrix Computations*. The Johns Hopkins University Press, 1983.

[gopal78]   K. Gopal, M. S. Nakhla, K. Singhal and J. Vlach. Distortion analysis of transistor networks. *IEEE Transactions on Circuits and Systems*, vol. CAS-25, no. 2, February 1978, pp. 99-106.

[grosz82]   Francis B. Grosz and Timothy N. Trick. Some modifications to Newton's method for the determination of the steady-state response of nonlinear oscillatory circuits. *IEEE Transactions on Computer-Aided Design of Integrated Circuits and Systems*, vol. CAD-1, no. 3, July 1982, pp. 116-120.

[gwarek74]   Wojciech K. Gwarek. *Nonlinear analysis of microwave mixers*. Masters thesis, Massachusetts Institute of Technology, September 1974.

[hachtel72]   Gary D. Hachtel. Vector and variability types in sparse matrix algorithms. In *Sparse Matrices and Their Applications*, Donald J. Rose and Ralph A. Willoughby (editors). Plenum Press, 1972, pp. 53-64.

[hale80]   Jack K. Hale. *Ordinary Differential Equations*. Krieger, 1980.

[hall76]   G. Hall and J. M. Watt (editors). *Modern Numerical Methods for Ordinary Differential Equations*. Oxford University Press, 1976.

[harris78]   Fredric J. Harris. On the use of windows for harmonic analysis with the discrete Fourier transform. *Proceedings of the IEEE*, vol. 66, no. 1, January 1978, pp. 51-83.

[held78]   Daniel N. Held and Anthony R. Kerr. Conversion loss and noise of microwave and millimeter-wave mixers. *IEEE Transactions on Microwave Theory and Techniques*, vol. MTT-26, no. 2, February 1978, pp. 49-61.

[hente86]   D. Hente and R. H. Jansen. Frequency domain continuation method for the analysis and stability investigation of nonlinear microwave circuits. *IEE Proceedings*, part H, vol. 133, no. 5, October 1986, pp. 351-362.

[hicks82a]   R. G. Hicks and P. J. Khan. Numerical analysis of nonlinear solid-state device excitation in microwave circuits. *IEEE Transactions on Microwave Theory and Techniques*, vol. MTT-30, no. 3, March 1982, pp. 251-259.

[hicks82b]   R. G. Hicks and P. J. Khan. Numerical analysis of subharmonic mixers using accurate and approximate models. *IEEE Transactions on Microwave Theory and Techniques*, vol. MTT-30, no. 12, December 1982, pp. 2113-2120.

[hirsch70]   Morris W. Hirsch and Stephen Smale. *Differential Equations, Dynamical Systems, and Linear Algebra.* Academic Press, 1970.

[ho75]   Chung-Wen Ho, Albert E. Ruehli and Pierce A. Brennan. The modified nodal approach to network analysis. *IEEE Transactions on Circuits and Systems*, vol. CAS-22, no. 6, June 1975, pp. 504-509.

[householder75]   Alston S. Householder. *The Theory of Matrices in Numerical Analysis.* Dover, 1975.

[huang89]   Tammy Huang. *Theoretical Aspects of Relaxation-Based and Nonlinear Frequency-Domain Circuit Simulation.* Ph. D. dissertation, University of California at Berkeley, April 1989. Available through Electronics Research Laboratory Publications, U. C. B., 94720, Memorandum No. UCB/ERL M89/51.

[kakizaki85]   Makiko Kakizaki and Tsutomu Sugawara. A modified Newton method for the steady-state analysis. *IEEE Transactions on Computer-Aided Design of Integrated Circuits and Systems,* vol. CAD-4, no. 4, October 1985, pp. 662-667.

[keller68]   Herbert B. Keller. *Numerical Methods for Two-Point Boundary Value Problems.* Ginn-Blaisdell, Waltham Mass., 1968.

[keller76]   Herbert B. Keller. *Numerical Solution of Two Point Boundary Value Problems.* Society for Industrial and Applied Mathematics, 1976.

[keller77]   Herbert B. Keller. Numerical solution of bifurcation and nonlinear eigenvalue problems. In *Applications of Bifurcation Theory,* Paul H. Rabinowitz (editor), 1977.

[kerr75]   A. R. Kerr. A technique for determining the local oscillator waveforms in a microwave mixer. *IEEE Transactions on Microwave Theory and Techniques,* vol. MTT-23, no. 10, October 1975, pp. 828-831.

[kerr79]   Anthony R. Kerr. Noise and loss in balanced and subharmonically pumped mixers. *IEEE Transactions on Microwave Theory and Techniques,* vol. MTT-27, no. 12, December 1979, pp. 938-950.

[kundert85]   Kenneth S. Kundert and Alberto Sangiovanni-Vincentelli. Nonlinear circuit simulation in the frequency-domain. In *ICCAD-85 Digest of Technical Papers.* IEEE International Conference on

Computer-Aided Design, November 1985.

[kundert86a]   Kenneth S. Kundert. Sparse matrix techniques. In *Circuit Analysis, Simulation and Design,* part 1, Albert E. Ruehli (editor). North-Holland, 1986, pp. 281-324.

[kundert86b]   Kenneth S. Kundert and Alberto Sangiovanni-Vincentelli. Simulation of nonlinear circuits in the frequency domain. *IEEE Transactions on Computer-Aided Design of Integrated Circuits and Systems,* vol. CAD-5, no. 4, October 1986, pp. 521-535. Also see comments by G. W. Rhyne and M. B. Steer in August 1989 issue of same journal (CAD 8-8, pp. 927-929).

[kundert88a]   K. Kundert, J. White and A. Sangiovanni-Vincentelli. An envelope-following method for the efficient transient simulation of switching power and filter circuits. In *ICCAD-88 Digest of Technical Papers.* IEEE International Conference on Computer-Aided Design, November 1988, pp. 446-449.

[kundert88b]   Kenneth S. Kundert, Alberto Sangiovanni-Vincentelli and Tsutomu Sugawara. Techniques for finding the periodic steady-state response of circuits. In *Analog Methods for Circuit Analysis and Diagnosis,* Takao Ozawa (editor). Marcel Dekker, 1988.

[kundert88c]   K. Kundert, J. White and A. Sangiovanni-Vincentelli. Mixed frequency-time approach for finding the steady-state solution of clocked analog circuits. *Proceedings of the IEEE 1988 Custom Integrated Circuits Conference,* May 1988, pp. 6.2.1-6.2.4.

[kundert88d]   Kenneth S. Kundert, Gregory B. Sorkin and Alberto Sangiovanni-Vincentelli. Applying harmonic balance to almost-periodic circuits. *IEEE Transactions on*

Microwave Theory and Techniques, vol. MTT-36, no. 2, February 1988, pp. 366-378.

[kundert89]   K. Kundert, J. White and A. Sangiovanni-Vincentelli. A mixed frequency-time approach for distortion analysis of switching filter circuits. *IEEE Journal of Solid-State Circuits*, vol. 24, no. 2, April 1989, pp. 443-451.

[kundert]   Kenneth S. Kundert. *Spectre: A Frequency-Domain Simulator for Nonlinear Circuits.* For information write to the EECS Industrial Liaison Program, University of California, Berkeley California, 94720.

[lewis87]   S. H. Lewis and P. R. Gray. A pipelined 5-Msample/s 9-bit analog-to-digital converter. *IEEE Journal of Solid-State Circuits*, vol. SC-22, December 1987, pp. 954-961.

[lipparini82]   Alessandro Lipparini, Ernesto Marazzi and Vittorio Rizzoli. A new approach to the computer-aided design of nonlinear networks and its application to microwave parametric frequency dividers. *IEEE Transactions on Microwave Theory and Techniques*, vol. MTT-30, no. 7, July 1982, pp. 1050-1058.

[luenberger84]   David G. Luenberger. *Linear and Nonlinear Programming.* Addison-Wesley, 1984.

[maas88]   Stephen A. Maas. *Nonlinear Microwave Circuits.* Artech House, 1988.

[mayaram88]   Kartikeya Mayaram and Donald O. Pederson. CODECS: A mixed-level device and circuit simulator. In *ICCAD-88 Digest of Technical Papers.* IEEE International Conference on Computer-Aided Design, November 1988, pp. 112-115.

[mccormick89]   Steven Paul McCormick. *Modeling and Simulation of VLSI Interconnections with Moments.* Ph. D. dissertation, Massachusetts Institute of Technology, June 1989.

[mees81]   A. I. Mees. *Dynamics of Feedback Systems.* Wiley, 1981.

[nagel75]   Laurence W. Nagel. *SPICE2: A Computer Program to Simulate Semiconductor Circuits.* Ph. D. dissertation, University of California at Berkeley, May 1975. Available through Electronics Research Laboratory Publications, U. C. B., 94720, Memorandum No. UCB/ERL M520.

[nakhla76]   M. S. Nakhla and J. Vlach. A piecewise harmonic balance technique for determination of the periodic response of nonlinear systems. *IEEE Transactions on Circuits and Systems,* vol. CAS-23, no. 2, February 1976, pp. 85-91.

[newton83]   Arthur Richard Newton and Alberto L. Sangiovanni-Vincentelli. Relaxation-based electrical simulation. *IEEE Transactions on Electron Devices,* vol. ED-30, no. 9, September 1983, pp. 1184-1207. (Also in *SIAM Journal on Scientific and Statistical Computing,* vol. 4, no. 3, September 1983 and *IEEE Transactions on Computer-Aided Design of Integrated Circuits and Systems,* vol. CAD-3, no. 4, October 1984).

[norin71]   Robert S. Norin and Christopher Pottle. Effective ordering of sparse matrices arising from nonlinear electrical networks. *IEEE Transactions on Circuit Theory,* vol. CT-18, no. 1, January 1971, pp. 139-145.

[orr86]   Jerry Orr. A stable 2-26.5 GHz two-stage dual-gate distributed MMIC amplifier. *1986 IEEE MTT-S International Microwave Symposium Digest,* June 1986.

[ortega70]   J. M. Ortega and W. C. Rheinboldt. *Iterative Solution of Nonlinear Equations in Several Variables.* Academic Press, 1970.

[parker89]   T. S. Parker and L. O. Chua. *Practical Numerical Algorithms for Chaotic Systems.* Springer-Verlag, 1989.

[penalosa83]   C. Camacho Penalosa. Numerical steady-state analysis of nonlinear microwave circuits with periodic excitation. *IEEE Transactions on Microwave Theory and Techniques*, vol. MTT-31, 1983, pp. 724-730.

[penalosa87a]   Carlos Camacho Penalosa and Colin S. Aitchison. Analysis and design of MESFET gate mixers. *IEEE Transactions on Microwave Theory and Techniques*, vol. MTT-35, no. 7, July 1987, pp. 643-652.

[penalosa87b]   C. Camacho Penalosa, L. Mariscal Rico and A. Alonso Pardo. Efficient calculation of partial derivatives in nonlinear conductances driven by periodic input signals. *Electronic Letters*, vol. 23, no. 11, 21 May 1987, pp. 565-566.

[petzold81]   Linda R. Petzold. An efficient numerical method for highly oscillatory ordinary differential equations. *SIAM Journal on Numerical Analysis*, vol. 18, no. 3, June 1981, pp. 455-479.

[press86]   William H. Press, Brian P. Flannery, Saul A. Teukolsky and William T. Vetterling. *Numerical Recipes: the Art of Scientific Computing.* Cambridge University Press, 1986.

[quarles89]   Thomas Quarles. *Analysis of Performance and Convergence Issues for Circuit Simulation.* Ph. D. dissertation, University of California at Berkeley, April 1989. Available through Electronics Research Laboratory Publications, U. C. B., 94720, Memorandum No. UCB/ERL M89/42.

[rabiner75]   Lawrence R. Rabiner and Bernard Gold. *Theory and Application of Digital Signal Processing.* Prentice-Hall, 1975.

[rabiner78]   Lawrence R. Rabiner and Ronald W. Schafer. *Digital Processing of Speech Signals.* Prentice-Hall, 1978.

[rhyne88]   George W. Rhyne, Michael B. Steer and Bevan D. Bates. Frequency-domain nonlinear circuit analysis using generalized power series. *IEEE Transactions on Microwave Theory and Techniques*, vol. MTT-36, no. 2, February 1988, pp. 379-387.

[rizzoli83]   Vittorio Rizzoli, Alessandro Lipparini and Ernesto Marazzi. A general-purpose program for nonlinear microwave circuit design. *IEEE Transactions on Microwave Theory and Techniques*, vol. MTT-31, no. 9, September 1983, pp. 762-770.

[rizzoli86]   V. Rizzoli, A. Lipparini and A. Neri. User-oriented software package for the analysis and optimisation of nonlinear microwave circuits. *IEE Proceedings,* part H, vol. 133, no. 5, October 1986, pp. 385-391.

[rizzoli87]   V. Rizzoli, C. Cecchetti and A. Lipparini. A general-purpose program for the analysis of non-linear microwave circuits under multitone excitation by multidimensional Fourier transform. In *1987 Proceedings of the 17th European Microwave Conference*, September 1987, pp. 635-640.

[rudin76]   Walter Rudin. *Principles of Mathematical Analysis.* McGraw-Hill, 1976.

[sangiovanni81] Alberto L. Sangiovanni-Vincentelli. Circuit simulation. In *Computer Design Aids for VLSI Circuits*, P. Antognetti, D. O. Pederson and H. De Man (editors). Sijthoff & Noordhoff, 1981, pp. 19-112.

[schuppert86]   Bernd Schuppert. A fast and reliable method for computer analysis of microwave mixers. *IEEE Transactions on Microwave Theory and Techniques*, vol. MTT-34, no. 1, January 1986, pp. 110-119.

Also see comments by M. E. Adamski in March 1987 issue of same journal (MTT 35-3, pp. 353).

[schutt-aine88]   Jose E. Schutt-Aine and Raj Mittra. Scattering parameter transient analysis of transmission lines loaded with nonlinear terminations. *IEEE Transactions on Microwave Theory and Technique*, vol. MTT-36, no. 3, March 1988, pp. 529-536.

[skelboe80]   Stig Skelboe. Computation of the periodic steady-state response of nonlinear networks by extrapolation methods. *IEEE Transactions on Circuits and Systems*, vol. CAS-27, no. 3, March 1980, pp. 161-175.

[skelboe82]   Stig Skelboe. Conditions for quadratic convergence of quick periodic steady-state methods. *IEEE Transactions on Circuits and Systems*, vol. CAS-29, no. 4, April 1982, pp. 234-239.

[smith87]   David A. Smith, William F. Ford and Avram Sidi. Extrapolation methods for vector sequences. *SIAM Review*, vol. 29, no. 2, June 1987, pp. 199-233.

[sorkin87]   Gregory B. Sorkin, Kenneth S. Kundert and Alberto Sangiovanni-Vincentelli. An almost-periodic Fourier transform for use with harmonic balance. In *1987 IEEE MTT-S International Microwave Symposium Digest*, vol. 2, June 1987, pp. 717-720.

[steer83]   Michael B. Steer and Peter J. Khan. An algebraic formula for the output of a system with large-signal, multifrequency excitation. *Proceedings of the IEEE*, vol. 71, no. 1, January 1983, pp. 177-179.

[stoer80]   J. Stoer and R. Bulirsch. *Introduction to Numerical Analysis*. Springer-Verlag, 1980.

[strang80]   Gilbert Strang. *Linear Algebra and Its Applications*. Academic Press, 1980.

[sugawara89]   T. Sugawara, M. Okumura and H. Tanimoto. An efficient small signal frequency analysis method for nonlinear circuits with two frequency excitations. *IEEE Transactions on Computer-Aided Design of Integrated Circuits and Systems*, to appear in 1989.

[ushida84]   A. Ushida and L. O. Chua. Frequency-domain analysis of nonlinear circuits driven by multi-tone signals. *IEEE Transactions on Circuits and Systems*, vol. CAS-31, no. 9, September 1984, pp. 766-778.

[ushida87]   A. Ushida, L. O. Chua and T. Sugawara. A substitution algorithm for solving nonlinear circuits with multi-frequency components. *International Journal on Circuit Theory and Application*, vol. 15, 1987, pp. 327-355.

[vidyasagar78]   M. Vidyasagar. *Nonlinear Systems Analysis*. Prentice-Hall, 1978.

[vlach83]   Jiri Vlach and Kishore Singhal. *Computer Methods for Circuit Analysis and Design*. Van Nostrand Reinhold, 1983.

[weeks73]   W. T. Weeks, A. J. Jimenez, G. W. Mahoney, D. Mehta, H. Qassemzadeh and T. R. Scott. Algorithms for ASTAP — a network analysis program. *IEEE Transactions on Circuit Theory*, vol. CT-20, no. 6, November 1973, pp. 628-634.

[white86]   Jacob K. White and Alberto Sangiovanni-Vincentelli. *Relaxation Techniques for the Simulation of VLSI Circuits*. Kluwer, 1986.

[yang82]   Ping Yang and Pallab K. Chatterjee. SPICE modeling for small geometry MOSFET circuits. *IEEE Transactions on Computer-Aided Design of Integrated Circuits and Systems*, vol. CAD-1, no. 4, October 1982, pp. 169-182.

Index

Definitions are denoted with bold page numbers, sections with §.